钱学森图书馆

錢學森

钱学森力学手稿

⑧

钱学森

西安交通大学出版社
XI'AN JIAOTONG UNIVERSITY PRESS

图书在版编目(CIP)数据
钱学森力学手稿. 8:英文/钱学森著. —西安:西安交通大学出版社,2013.2
ISBN 978-7-5605-4532-5

Ⅰ.①钱…　Ⅱ.①钱…　Ⅲ.①钱学森(1911～2009)-力学-手稿-英文　Ⅳ.①O3-53

中国版本图书馆 CIP 数据核字(2012)第 206467 号

书　　名　钱学森力学手稿 8
著　　者　钱学森
责任编辑　王　欣

出版发行　西安交通大学出版社
(西安市兴庆南路 10 号　邮政编码 710049)
网　　址　http://www.xjtupress.com
电　　话　(029)82668357　82667874(发行中心)
(029)82668315　82669096(总编办)
传　　真　(029)82668280
印　　刷　中煤地西安地图制印有限公司

开　　本　787mm×1092mm　1/16　**印张** 11.25　**字数** 271 千字
版次印次　2013 年 1 月第 1 版　2013 年 1 月第 1 次印刷
书　　号　ISBN 978-7-5605-4532-5/O·407
定　　价　70.00 元

读者购书、书店添货、如发现印装质量问题,请与本社发行中心联系、调换。
订购热线:(029)82665248　(029)82665249
投稿热线:(029)82664954
读者信箱:jdlgy@yahoo.cn

出 版 前 言

2011年12月11日是西安交通大学杰出校友钱学森先生的百年诞辰。为缅怀钱学森学长，学习他的科学思想和卓越风范，展示其丰功伟绩和人格魅力，西安交通大学举办了"纪念钱学森诞辰100周年"系列活动：作为制片方之一，参与西部电影集团摄制传记故事片《钱学森》；与中央电视台合作，出品纪录片《实验班的故事——沿着钱学森走过的路》；扩建钱学森生平业绩展馆，向校内外开放；举办钱学森科学与教育思想研讨会；出版发行《钱学森力学手稿》、《钱学森年谱（初编）》、《钱学森第六次产业革命思想探微丛书》等。

钱学森先生在美国深造和工作期间留下大量珍贵手稿，这些手稿真实展示了钱学森先生博大精深的学识、开拓求实的精神和严谨奋进的作风，是钱老勇攀科学高峰和严谨治学的集中体现。这里，我们将部分原稿整理汇集成册，出版《钱学森力学手稿》，作为钱老百年诞辰的献礼。

《钱学森力学手稿》共10卷，包含两部分内容。第一部分是草稿，包括扁壳、球壳和圆柱壳屈曲分析的公式推导和数值演算。在研究圆柱壳轴压屈曲问题时，为了求得圆柱壳体的临界压力，在有关的五百多页草稿中，对多达二十多种可能的屈曲模态逐一进行公式推演和数值计算，最终才找到满意的并在论文中采用的屈曲模态。仔细观察草稿中的数据列表，每个数字有效位数都长达八位，在手摇机械式计算机作为主要计算工具的年代，这串串数字凝聚着多少现今难以想象的艰辛劳动。

第二部分是手稿，以航空航天工程为核心，涵盖空气动力学、固体力学、火箭技术、工程控制论和物理力学等领域的部分学术论文手稿、打印稿和讲义。

《钱学森力学手稿》是在西安交通大学校领导的大力支持下，由西安交通大学航天航空学院沈亚鹏教授整理完成。图书出版过程中得到了西安交通大学党委宣传部、校友关系发展部、图书馆、航天航空学院等的积极协助，在此深表感谢。

Contents

Section 1

Two Dimensional Subsonic Flow of Compressible Fluids

2 Copies

Summary

The basic concept of the present paper is to use ~~the~~ a tangent line to the adiabatic pressure-volume curve as an approximation to the curve itself. First, the general characteristics of such ~~an~~ a fluid are shown. Then in Section I, a theory is developed which in ~~main features~~ principle is similar to that of Demtchenko and Busemann but is more general and can be applied to flow with velocity approach~~es~~ing that of sound.]] The theory is then applied to calculate the flow over elliptic cylinders. In Section II the work of H. Bateman is applied to this approximate~~y~~ adiabatic fluid and ~~two equations are given which express the relations between the velocity and the pressure distributions over a body in compressible flow to the velocity and the pressure distributions over the same body in incompressible flow. This These equations are then used to predict the high speed characteristics of airfoils.~~ ~~and result~~ the results obtained are essentially the same as ~~that~~ those obtained in Section I.

]] The theory is put into a form, by which, knowing the incompressible flow over a body, the compressible flow over a similar body ~~near near to the first one in incompressible flow~~ can be calculated.

TWO-DIMENSIONAL SUBSONIC FLOW

OF COMPRESSIBLE FLUIDS

by

Hsue-shen Tsien
California Institute of Technology

Introduction

Assuming that the pressure is a single-valued function of density only, the equations of two-dimensional irrotational motion of compressible fluids can be reduced to a single non-linear equation of the velocity potential. In the supersonic case, that is, in the case when ~~the~~ flow velocity is everywhere greater than that of local sound velocity, the problem is solved by Meyer & Prandtl and Busemann using the method of characteristics. The essential difficulty of this problem lies in the subsonic case, that is, in the case when ~~the~~ flow velocity is everywhere smaller than but near ~~to~~ the local sound velocity, because then the method of characteristics cannot be used. Glauert & Prandtl (Ref. 1) treated the case when the disturbance to the parallel rectilinear flow due to presence of a solid body is small. They were then able to linearize the differential equation for the velocity potential and obtained an equation very similar to ~~those~~ that for ~~the~~ incompressible fluids. But there are usually stagnation points either in the surface of the body or in the field of flow, where the disturbance is no longer small. Hence, it is doubtful whether the linear theory can be applied to the flow near a stagnation point. On the same ground, the theory breaks down in the case of bodies whose dimension across the stream is not small compared with the dimension parallel to the stream.

To treat these cases ~~in which the body is blunt-nosed~~, Janzen and Rayleigh developed the method of successive approximations. This method ~~was explained physically and~~ was put into a more convenient form by L. Poggi ~~(Ref. 2)~~ and P.A. Walther ~~(Ref. 3)~~. Recently C. Kaplan (Ref. 2) treated the case of flow over Joukowsky airfoils and elliptic cylinders using Poggi's method. However, the method is rather tedious and the convergence very slow if the local velocity of sound is approached.

Begin Molenbroek ~~(Ref. 5)~~ and Tschapligin ~~(Ref. 6)~~ suggested the use of the magnitude of velocity w and inclination β of velocity to a chosen axis as independent variables, and were thus able to reduce the equation of velocity potential to a linear equation. This equation was solved by Tschapligin. ~~(Ref. 6) and recently put into a more convenient form by F. Clauser and M. Clauser (Ref. 7)~~. The solution is essentially a series each term of which is a product of a hypergeometric function of w and a trigonometric function of β. The main difficulty in practical application of this solution is to obtain a proper set of boundary conditions in the plane of independent variables w, β and to put the solution in a closed form.

Tschapligin ~~(Ref. 6)~~ has shown that a great simplification of the equation in the hodograph plane results if the ratio of the specific heats of the gas is equal to -1. Since all real gases have their ratio of specific heats between 1 and 2, the value -1 seems without practical significance. It was Demtchenko (Ref. 3) and Busemann (Ref. 4) who made clear clarified the meaning of this specific value of -1. They found that this really means to take the tangent of the pressure-volume curve as an approximation to the curve itself. However, they limit themselves to use the tangent at the state of ~~rest of~~ the gas, corresponding to the stagnation point of flow. Thus their theory can only be applied to a flow with velocities up to about one half ~~that of sound velocity~~ [illegible]. Recently, during a personal discussion, Th. von Kármán suggested to the author that the theory can be generalized ~~to~~ use the tangent at the state of [the] gas corresponding to ~~the~~ undisturbed parallel flow. Thus the range of usefulness of the theory can be greatly extended. This is carried out in the first ~~part~~ [section] of the present paper.

~~In the second part of this paper~~ [then] this theory, based upon Demtschenko and Busemann's work, is applied to the case of flow over elliptic cylinders and the results compared with those of S.G. Hooker (Ref. ~~10~~ [5]) and C. Kaplan (Ref. ~~4~~ [2]). Furthermore, results calculated by Glauert-Prandtl linear theory are also included for comparison.

Recently, H. Bateman (Ref. ~~11~~ [6]) demonstrated a remarkable reciprocity of two fields of flow of two fluids related by a certain point transformation. It will be shown in the ~~third part~~ [second section] of this paper that the flow of [an] incompressible fluid and the flow of compressible fluid approximated by the use of [the] tangent to adiabatic pressure-volume curve can be interpreted as such a point transformation. It is thus possible to obtain a solution for compressible flow whenever a solution of incompressible flow is known. ~~The difficulty, however, lies in~~

This transformation from a flow of incompressible fluid to a flow of compressible fluid is found, however, essentially to be the same as that developed ~~in Section I~~ from Demtchenko's and Busemann's work. ~~The only difference~~ is

~~body at any Mach's number whenever the low speed characteristics of the flow over the same body are known. The characteristics of the~~

incompressible flow can either be obtained by the well-known method of conformal mapping or by experiments. Due to the fact that practical aerodynamic engineers usually have the low speed characteristics at hand and that high speed data have to be obtained by use of a costly high speed wind tunnel the above mentioned relations are believed to be of considerable use to them.

~~In the fourth part of the paper,~~ ~~the~~ this theory ~~developed in Part III~~ is ~~applied~~ then used to correlate airfoil data obtained by J. Stack (Ref. ~~12~~ 7) in the N.A.C.A. 24" high speed wind tunnel. The agreement with theory is found to be satisfactory. Then this theory is applied to predict the compressibility effect on the lift and moment of N.A.C.A. 4412 airfoil using experimentally determined pressure distribution over the same airfoil at low speed. The result is again compared with the more simple Glauert-Prandtl theory.

Approximation to the Adiabatic Relation

~~Approximation to Adiabatic~~

If p is the pressure, v is the specific volume and γ is the ratio of specific heats of a gas, the adiabatic relation $pv^{-\gamma}=$ constant is a curve in the p-v plane as shown in Fig. 1a. ~~The~~ The conditions near the point (p_1, v_1) which corresponds to ~~the~~ a state of ~~fluid at~~ undisturbed flow can be approximated by the tangent to the curve at that point. The equation of the tangent at this point can be written as:

$$p_1 - p = C(v_1 - v) = C(\rho_1^{-1} - \rho^{-1}) \tag{1}$$

a state of undisturbed ~~fluid~~ flow —

where C is the slope of the tangent and ρ is the density of the fluid. The slope C must be equal to the slope of the curve at the point (p_1, v_1), therefore,

$$C = \left(\frac{dp}{dv}\right)_1 = \left(\frac{dp}{d\rho}\frac{d\rho}{dv}\right)_1 = -\left(\frac{dp}{d\rho}\right)_1 \rho_1^2 = -a_1^2\rho_1^2$$

where a_1 is the sound velocity corresponding to the conditions p_1, v_1. Thus eq. (1) can be written as:

$$p_1 - p = a_1^2\rho_1^2\left(\frac{1}{\rho} - \frac{1}{\rho_1}\right) \qquad (2)$$

This is an approximation pressure-density relation, and is shown in Fig. 1 together with the true adiabatic relation.

The generalized Bernoulli theorem for compressible fluids is

$$\frac{1}{2}w_2^2 - \frac{1}{2}w_3^2 = \int_2^3 \frac{dp}{\rho} \qquad (3)$$

where w is the velocity of the gas and the subscripts 2 and 3 denote two different states of the fluid. By substituting eq. (2) into eq. (3), the following relation is obtained:

$$\frac{1}{2}w_2^2 - \frac{1}{2}w_3^2 = \frac{1}{2}a_1^2\rho_1^2\left\{\frac{1}{\rho_2^2} - \frac{1}{\rho_3^2}\right\} \qquad (4)$$

Now if $w_3 = 0$, $w_2 = w$, $\rho_3 = \rho_0$, and $\rho_2 = \rho$, with the subscript 0 denoting the state of the fluid corresponding to the stagnation point of flow, eq. (4) gives:

$$\frac{a_1^2\rho_1^2}{\rho_0^2} + w^2 = \frac{\rho_1^2 a_1^2}{\rho^2} \qquad (5)$$

If the square of sound velocity a_{II}^2 is defined (as is usually done) as the derivative of p with respect to ρ, eq. (2) gives:

$$a^2\rho^2 = \frac{dp}{d\rho}\rho^2 = a_1^2\rho_1^2 = \text{constant} \tag{6}$$

Therefore, eq. (5) can be written as:

$$\left(\frac{\rho}{\rho_0}\right)^2 = 1 - \frac{w^2}{a^2} \tag{7}$$

Similarly,

$$\left(\frac{\rho_0}{\rho}\right)^2 = 1 + \frac{w^2}{a_0^2} \tag{8}$$

It is interesting to ~~notice~~ note that from eq. (8) the density decreases as the velocity increases, as is maybe expected. Thus eq. (6) ~~shows~~ indicates that the local velocity of sound increases as the velocity increases. This is just opposite to ~~the~~ a real gas, because in the case of an adiabatic flow of a real gas it is well known that the temperature of the gas decreases as the velocity [~~of gas~~ the] ~~is increased, and~~ increases thus the local sound velocity, [being proportional to the square root of the temperature] also decreases. However, in the present approximate theory, the ratio $\frac{w}{a}$ or Mach's number still increases as the velocity increases, as can be seen by eq. (7). But this ratio only reaches the value unity when $\rho = 0$, or from eq. (8) when $w = \infty$. It is thus seen that the entire regime of flow is subsonic and thus the differential equation of the velocity potential [is always of an elliptic type, that is, always of same the type as the differential equation of the velocity potential of incompressible fluids.] ~~of an incompressible~~. This is the reason why the complex representation of the velocity potential and the stream function is possible for all cases, as will be shown in the following paragraphs. However, one should realize that the portion of the tangent that could be used as an approximation to the true adiabatic relation is that portion which lies in the first quadrant. Thus the upper limit ~~of~~ velocity for practical application of the theory ~~is~~ occurs

~~when~~ at $p=0$. By using eqs. (17) and (18), this upper limit is found to be

$$\left(\frac{w}{w_1}\right)_{max} = \frac{1}{\left(\frac{w_1}{a_1}\right)} \sqrt{\left(\frac{p_1}{a_1^2 \rho_1}+1\right)^2 - \left\{1-\left(\frac{w_1}{a_1}\right)^2\right\}} \quad (9)$$

Since the point (p_1, ρ_1)] being the tangent point to the true adiabatic curve, lies on the ~~true adiabati~~c curve, the relation $a_1^2 = \gamma \frac{p_1}{\rho_1}$]] which is true for the adiabatic relation $p\rho^{-\gamma}=$ constant can be used, and eq. (9) becomes:

$$\left(\frac{w}{w_1}\right)_{max} = \frac{1}{\left(\frac{w_1}{a_1}\right)} \sqrt{\left(\frac{1}{\gamma}+1\right)^2 - \left\{1-\left(\frac{w_1}{a_1}\right)^2\right\}} \quad (10)$$

This relation is plotted in Fig. 2 with $\gamma = 1.405$. Since for most practical cases it is not likely that the ratio $\left(\frac{w}{w_1}\right)$ will rise to values much higher than 2, p will remain positive, and this theory will be sufficient to give an approximate solution.

Section II ~~General Theory~~ Hodograph Method

If the flow is irrotational, there exists a velocity potential ϕ such that

$$\frac{\partial \phi}{\partial x} = u, \qquad \frac{\partial \phi}{\partial y} = v \quad (11)$$

where u, v are the components of w in the x and y direction, respectively. The equation of continuity,

$$\frac{\partial}{\partial x}\left(\frac{\rho}{\rho_0} u\right) + \frac{\partial}{\partial y}\left(\frac{\rho}{\rho_0} v\right) = 0$$

will be satisfied, if the stream function ψ is introduced such that

$$\frac{\rho}{\rho_0} u = \frac{\partial \psi}{\partial y}, \qquad -\frac{\rho}{\rho_0} v = \frac{\partial \psi}{\partial x} \quad (12)$$

Now if the angle of inclination of the velocity w to the x axis is β , eqs. (11) and (12) give:

$$d\phi = w\cos\beta\, dx + w\sin\beta\, dy$$
$$d\psi = -w\frac{\rho}{\rho_0}\sin\beta\, dx + w\frac{\rho}{\rho_0}\cos\beta\, dy \qquad (13)$$

Solving for dx and dy ,

$$dx = \frac{\cos\beta}{w}d\phi - \frac{\sin\beta}{w}\frac{\rho_0}{\rho}d\psi$$
$$dy = \frac{\sin\beta}{w}d\phi + \frac{\cos\beta}{w}\frac{\rho_0}{\rho}d\psi \qquad (14)$$

So long as the correspondence between the physical ~~plane~~ and hodograph plane is one to one, or mathematically $\frac{\partial(x,y)}{\partial(u,v)} \neq 0$, ~~one can~~ can be expressed x and y as functions of w , β, and ~~as~~ ϕ and ψ as functions of w, β . Thus,

$$d\phi = \phi'_w\, dw + \phi'_\beta\, d\beta \qquad (15)$$
$$d\psi = \psi'_w\, dw + \psi'_\beta\, d\beta$$

where primes indicate the derivative with respect to the independent variables indicated as subscripts. Now substituting eq. (15) into eq. (14), ~~one has~~ the following expressions for dx & dy are obtained:

$$dx = \left(\frac{\cos\beta}{w}\phi'_w - \frac{\sin\beta}{w}\frac{\rho_0}{\rho}\psi'_w\right)dw + \left(\frac{\cos\beta}{w}\phi'_\beta - \frac{\sin\beta}{w}\frac{\rho_0}{\rho}\psi'_\beta\right)d\beta$$

$$dy = \left(\frac{\sin\beta}{w}\phi'_w + \frac{\cos\beta}{w}\frac{\rho_0}{\rho}\psi'_w\right)dw + \left(\frac{\sin\beta}{w}\phi'_\beta + \frac{\cos\beta}{w}\frac{\rho_0}{\rho}\psi'_\beta\right)d\beta \qquad (16)$$

Since the left-hand side of eqs.(16) are exact differentials, ~~one can apply~~ the reciprocity relation can be applied and ~~obtain~~ therefore :

$$\frac{\partial}{\partial\beta}\left(\frac{\cos\beta}{w}\phi'_w - \frac{\sin\beta}{w}\frac{\rho_0}{\rho}\psi'_w\right) = \frac{\partial}{\partial w}\left(\frac{\cos\beta}{w}\phi'_\beta - \frac{\sin\beta}{w}\frac{\rho_0}{\rho}\psi'_\beta\right)$$
$$\frac{\partial}{\partial\beta}\left(\frac{\sin\beta}{w}\phi'_w + \frac{\cos\beta}{w}\frac{\rho_0}{\rho}\psi'_w\right) = \frac{\partial}{\partial w}\left(\frac{\sin\beta}{w}\phi'_\beta + \frac{\cos\beta}{w}\frac{\rho_0}{\rho}\psi'_\beta\right) \qquad (17)$$

Carrying out these differentiations and simplifying with the aid of eq. (7), eq. (17) gives:

$$\begin{aligned} -\frac{\sin\beta}{w}\phi'_w + \frac{\cos\beta}{w}\frac{\rho_0}{\rho}\psi'_w &= -\frac{\cos\beta}{w^2}\phi'_\beta + \frac{\sin\beta}{w^2}\frac{\rho}{\rho_0}\psi'_\beta \\ \frac{\cos\beta}{w}\phi'_w - \frac{\sin\beta}{w}\frac{\rho_0}{\rho}\psi'_w &= -\frac{\sin\beta}{w^2}\phi'_\beta - \frac{\cos\beta}{w^2}\frac{\rho}{\rho_0}\psi'_\beta \end{aligned} \tag{18}$$

Solving for ϕ'_w and ψ'_β,

$$\begin{aligned} \phi'_w &= -\frac{\rho}{\rho_0}\frac{1}{w}\psi'_\beta \\ \phi'_\beta &= \frac{\rho_0}{\rho} w \psi'_w \end{aligned} \tag{19}$$

Eq. (19) can be further simplified by introducing a new variable ω, such that

$$d\omega = \frac{\rho}{\rho_0}\frac{dw}{w} \tag{20}$$

Then eq. (19) becomes:

$$\begin{aligned} \phi'_\omega &= -\psi'_\beta \\ \phi'_\beta &= \psi'_\omega \end{aligned} \tag{21}$$

This can be easily recognized as the Riemann-Cauchy differential equation, and thus $\phi + i\psi$ must be an analytic function of $\omega - i\beta$. However, for convenience of calculation, another new set of independent variables $U = W\cos\beta$, $V = W\sin\beta$ are introduced where $W = a_0 e^{\omega}$

Then eq. (21) can be written as:

$$\begin{aligned} \frac{\partial\phi}{\partial U} &= \frac{\partial\psi}{\partial(-V)} \\ \frac{\partial\phi}{\partial(-V)} &= -\frac{\partial\psi}{\partial U} \end{aligned} \tag{22}$$

By integrating eq. (20),

$$W = \frac{2a_0 w}{\sqrt{a_0^2 + w^2} + a_0} \tag{23}$$

and

$$w = \frac{4a_0^2 W}{4a_0^2 - W^2} \tag{24}$$

Substituting into eq. (8), ~~we have~~ the following expression for the density ratio $\frac{\rho_0}{\rho}$ is obtained

$$\frac{\rho_0}{\rho} = \frac{4a_0^2 + W^2}{4a_0^2 - W^2} \tag{25}$$

Eqs. (22), (23), (24) and (25) are the basic equations of the present theory. One recognizes eq. (22) as the Riemann-Cauchy differential equation, and thus the complex potential $F = \phi + i\psi$ must be an analytic function of $\bar{W} = U - iV$, or :

$$\begin{aligned} \phi + i\psi &= F(U - iV) = F(\bar{W}) \\ \phi - i\psi &= \bar{F}(U + iV) = \bar{F}(W) \end{aligned} \tag{26}$$

In eq. (26), $\bar{W}$, and $\bar{F}$, are the complex conjugates of W & F respectively.

Now, it is necessary to find the values of x and y ~~corresponds~~ corresponding to a given set of values of U and V, i.e., to find the transformation from the hodograph plane to the physical plane. By using eqs. (24) and (25), eq. (14) can be written as:

$$\begin{aligned} dx &= \frac{U \cdot d\phi}{W^2}\left\{1 - \frac{W^2}{4a_0^2}\right\} - \frac{V \cdot d\psi}{W^2}\left\{1 + \frac{W^2}{4a_0^2}\right\} \\ dy &= \frac{V \cdot d\phi}{W^2}\left\{1 - \frac{W^2}{4a_0^2}\right\} + \frac{U \cdot d\psi}{W^2}\left\{1 + \frac{W^2}{4a_0^2}\right\} \end{aligned} \tag{27}$$

where $W^2 = U^2 + V^2$. These equations can be combined into one equation by means of eq. (26). Thus,

$$dz = dx + i\,dy = \frac{dF}{\bar{W}} - \frac{W \cdot d\bar{F}}{4a_0^2} \tag{28}$$

Hence, if an analytic function $F(\bar{W})$ is given for each value of W, the corresponding real velocity w can be calculated by eq. (24). Then the coordinate of the point in the physical plane at which this velocity occurs can be calculated by means of integrating eq. (28). The pressure at this point is then given by eq. (2). However, there is an objection to this procedure. That is, the investigator is unable to predict ~~has no way of telling~~ whether ~~with~~ the chosen function $F(\bar{w})$, ~~he is going to obtain~~ will give the desired shape of the solid boundary and flow pattern. ~~[illegible]~~ In other words, this procedure still suffers the difficulty of boundary conditions as ~~it~~ is common to all hodograph methods.

Transformation from Incompressible Flow to Compressible Flow ~~with Similar Boundaries~~ Approximately

However, due to the particular simple relation of eq. (28), ~~one~~ can be ascertained approximately the resulting shape of the body, by starting with the ~~particular~~ function,

$$F(\bar{w}) = \phi + i\psi = W_1\, G(\zeta) \qquad (29)$$

where W_1 is the transformed undisturbed velocity to be interpreted by eq. (23), and ζ is the complex coordinate $\xi + i\eta$. This function is so chosen as to give the flow of an incompressible fluid over the desired body shape in coordinates ξ and η. ~~Now interpreting~~ The real velocity in the $\xi\eta$ plane of the incompressible fluid is interpreted as the transformed velocity W in the hodograph plane for the compressible fluid, ~~Then~~ it is known that

$$\bar{W} = W_1 \frac{dG(\zeta)}{d\zeta} \qquad (30)$$

Thus,

$$W = W_1 \frac{d\bar{G}(\bar{\zeta})}{d\bar{\zeta}} \qquad (31)$$

where $\bar{G}$ & $\bar{\zeta}$ are the complex conjugates of G & ζ respectively.

~~where $\bar{G}$, $\bar{W}$ are is the complex conjugate of G.~~

With ~~Substituting~~ eqs. (30) and (31), ~~into~~ eq. (28), ~~one has~~ gives:

$$dz = d\zeta - \lambda_1 \left[\frac{d\bar{G}(\bar{\zeta})}{d\bar{\zeta}}\right]^2 d\bar{\zeta}$$

where $\lambda = \frac{1}{4}\left(\frac{W_1}{a_0}\right)^2$

Integrating,

$$z = \zeta - \lambda \int \left(\frac{d\bar{G}}{d\bar{\zeta}}\right)^2 d\bar{\zeta} \qquad (32)$$

Therefore, the complex coordinate in the physical plane of the compressible fluid is equal to the corresponding complex coordinate in the physical plane of the incompressible fluid plus a correction term. Since this correction term is usually small, the resulting shape of the body will be quite similar to the one in the incompressible fluid. The factor λ in the correction term depends upon Mach's number of the undisturbed flow only. This can be shown by means of eqs. (7), (8), and (23), because from those equations the following relation is obtained:

$$\lambda = \frac{1}{4}\left(\frac{W_1}{a_0}\right)^2 = \frac{\left(\frac{w_1}{a_1}\right)^2}{\left\{1+\sqrt{1-\left(\frac{w_1}{a_1}\right)^2}\right\}^2} \qquad (33)$$

where $\left(\frac{w_1}{a_1}\right)$ is Mach's number of the undisturbed flow. The values of λ for different values of Mach's number $\left(\frac{w_1}{a_1}\right)$ are plotted in fig. 3. To calculate the velocity in the physical plane, W is first obtained from eq. (30) and then with eq. (23):

$$\frac{w}{w_1} = \left\{1-\lambda\right\} \frac{\frac{|W|}{W_1}}{1-\lambda\left\{\frac{|W|}{W_1}\right\}^2} \qquad (34)$$

With little calculation the following expression for pressure coefficient ϖ at any point is defined as $\varpi = \frac{p-p_0}{\frac{1}{2}\rho_1 w_1^2}$, then by using eq. (2) the following relation is obtained

$$\varpi = (1+\lambda)\frac{1-\left(\frac{|W|}{W_1}\right)^2}{1-\lambda\left(\frac{|W|}{W_1}\right)^2} \qquad (35)$$

Flow over ~~an~~ Elliptic Cylinders

~~In this part,~~ the theory ~~developed~~ ~~in Part I~~ will now be applied to calculate the flow over an elliptic cylinder at zero angle of attack. The incompressible flow over an elliptic cylinder in the complex coordinate ζ can be obtained by applying ~~the~~ Joukowsky's transformation to the flow over a circular cylinder in the complex coordinate ζ' with ~~its~~ the center of the circle located at the origin of the ζ' plane. Therefore, the function $F(\overline{W})$ or $W_1 G(\zeta)$ can be written as:

$$\left.\begin{aligned} F &= W_1\left[\zeta' + \frac{a^2}{\zeta'}\right] \\ \bar{F} &= \bar{W}_1\left[\bar{\zeta}' + \frac{a^2}{\bar{\zeta}'}\right] \end{aligned}\right\} \qquad (36)$$

~~and~~ with

$$\zeta = \zeta' + \frac{1}{\zeta'} \qquad (37)$$

It is convenient to carry out the computation by using the ζ' coordinates. Thus eq. (32) is rewritten in the following form:

$$z = \left(\zeta' + \frac{1}{\zeta'}\right) - \frac{1}{4}\left(\frac{W_1}{a_0}\right)^2 \int \left(\frac{dG}{d\zeta'}\right)^2 \frac{d\bar{\zeta}'}{\frac{d\bar{\zeta}}{d\bar{\zeta}'}} \qquad (38)$$

If ~~one limits one's self to calculate the~~ only the conditions over the surface of the elliptic cylinder ~~only.~~ are concerned, then

$$\begin{aligned} \zeta' &= b\,e^{i\theta} \\ \bar{\zeta}' &= b\,e^{-i\theta} \end{aligned} \qquad (39)$$

where θ is the argument as shown in Fig. ~~4~~ and b is the radius of the circular section in the ζ' - plane which determines the thickness ratio of the elliptic section in the ζ - plane. Substituting eqs. (36), (37) and (39) into eq. (38), and carrying out the integration, the following expressions for the x and y coordinates corresponding to ζ' are obtained by separating the real and imaginary parts,

$$x = \left(b + \frac{1}{b}\right)\cos\theta - \lambda_1 \left\{ b(1+b^2)\cos\theta + \frac{(b^2-1)^2}{4}\log\frac{(b^2-1)^2 + 4b^2\sin^2\theta}{(b^2+2b\cos\theta+1)} \right\}$$

$$y = \left(b - \frac{1}{b}\right)\sin\theta + \lambda_1 \left\{ b(1-b^2)\sin\theta + \frac{(b^2-1)^2}{2}\tan^{-1}\frac{2b\sin\theta}{b^2-1} \right\} \qquad (40)$$

The horizontal and vertical semi-axis of the ~~near~~ approximately elliptic section can then be calculated by substituting $\theta = 0$ and $\theta = \frac{\pi}{2}$ respectively into eq. (40). The thickness ratio δ is thus obtained as:

$$\delta = \left(\frac{b^2-1}{b^2+1}\right)\left\{ \frac{1 + \lambda_1\left[-b^2 + \frac{b(b^2-1)}{2}\tan^{-1}\frac{2b}{b^2-1}\right]}{1 - \lambda_1\left[b^2 + \frac{b(b^2-1)}{2}\left(\frac{b^2-1}{b^2+1}\right)\log\left(\frac{b-1}{b+1}\right)\right]} \right\} \qquad (41)$$

For a given thickness ratio and Mach's number ~~of~~ for undisturbed ~~stream~~ flow, the value of λ_1 is first computed by means of eq. (33), and then eq. (41) is solved graphically for b.

After b is obtained, the coordinate x and y for each value of θ can be computed by using eq. (40). It is fortunate that the values of x, y so obtained lies very close to the true elliptic section ~~as shown in Fig. 5~~. Hence, the velocity and the pressure distribution obtained by using eqs. (34) and (35) are considered as those over the true elliptic sections.

Calculations for two thickness ratio, $\delta = 0.5$ and $\delta = 0.1$, are carried out and the results shown in Fig. 5 & 6, together with those of Kaplan (Ref. 2). Hooker's results (Ref. 5) are very close to those of Kaplan. Computations are also carried out using the more simple theory of Glauert and Prandtl (Ref 1) and the results are included in Fig. 5 & Fig. 6 in order to compare with those of Kaplan and the present theory.

¶ The difference between the various theories lies in the assumptions which are made to simplify the calculations. Glauert & Prandtl assumed that the disturbance introduced by the solid body to the parallel flow is small. In other words, they treated the flow over a body of small thickness ratio. On the other hand, Kaplan & Hooker assumed that the Mach's number of the undisturbed flow is small, so that terms containing the third & higher powers of Mach's number can be neglected. The present theory is essentially an improvement of the Glauert-Prandtl theory so that the effect of large disturbances to the parallel flow is taken into account approximately. Therefore for flow over thin sections at high Mach's number, the result of the present theory should agree well with the Glauert-Prandtl theory, especially at points not too close to the stagnation point. The results of Kaplan & Hooker should show smaller effect of the compressibility due to their second order approximation.

For flow over thick sections at lower Mach's number, the situation is reversed. In this case, results of the present theory should give better agreement with the results obtained by Kaplan & Hooker than ~~those~~ with those obtained from the Glauert - Prandtl theory. The above reasoning is substantiated by Fig. 5 & 6.

Critical Velocities for Elliptic Cylinders

If the velocity of flow over a body is gradually increased, the maximum local velocity in the field will also be increased. When a ~~certain velocity of the undisturbed stream is reached~~ flow the maximum local velocity reaches the local velocity of sound, ~~Then~~ shock waves ~~suddenly~~ appear and the drag of the body suddenly increases. ~~Therefore~~, This velocity is therefore of considerable interest to practical engineers and is usually referred to as the critical velocity of the body. It is shown by Kaplan (Ref. 2) and others that at this critical condition the ratio of maximum velocity w_{max} in the field to that of the undisturbed velocity w_1, is related to Mach's number $\frac{w_1}{a_1}$ of the undisturbed fluid in the following manner:

$$\left(\frac{w_{max}}{w_1}\right) = \left\{ \frac{2}{\gamma+1} \frac{1}{\left(\frac{w_1}{a_1}\right)^2} + \frac{\gamma-1}{\gamma+1} \right\}^{\frac{1}{2}} \tag{42}$$

w_{max} in the flow over an elliptic cylinder at zero angle of attack occurs at the top of the cylinder. Using eqs. (34) and (33) the value of $\frac{w_{max}}{w_1}$ is found to be:

$$\frac{w}{w_1} = \frac{\left(\frac{2b^2}{b^2+1}\right)^2}{1 + \frac{\left[1-\left(\frac{2b^2}{b^2+1}\right)^2\right]\left(\frac{w_1}{a_1}\right)^2}{2\left(1+\sqrt{1-\left(\frac{w_1}{a_1}\right)^2}\right)\sqrt{1-\left(\frac{w_1}{a_1}\right)^2}}} \tag{43}$$

Equating eqs. (42) and (43), the equation for calculating the critical Mach's number $\left(\frac{w_1}{a_1}\right)$ of the undisturbed flow for each value of b is:

$$\frac{2}{\gamma+1} \frac{1}{\left(\frac{w_1}{a_1}\right)_{cr}} + \frac{\gamma-1}{\gamma+1} = \frac{\left(\frac{2b^2}{b^2+1}\right)^2}{1 + \frac{\left[1-\left(\frac{2b^2}{b^2+1}\right)^2\right]\left(\frac{w_1}{a_1}\right)^2_{cr}}{2\left(1+\sqrt{1-\left(\frac{w_1}{a}\right)^2_{cr}}\right)\sqrt{1-\left(\frac{w_1}{a_1}\right)^2_{cr}}}} \tag{44}$$

Knowing $\left(\frac{w_1}{a_1}\right)_{cr}$ for each value of b, the corresponding value of δ can be calculated by means of eqs. (34) and (41). Fig. 7 shows the result of this calculation with Kaplan's

lower value of the critical Mach's number

value included for comparison. It is seen that the critical Mach's number is lower than ~~those~~ that obtained by Kaplan. This ~~illustrates stronger~~ indicates a more pronounced effect of compressibility of a fluid ~~shown by ... theory~~ and is consistent with the results shown in Figs. 5 6.

Section II
The Use of ~~III.~~ Lift + Drag Functions

If two new functions X and Y are defined by

$$p_0\, dX = p\, dy + \rho_0 u\, d\psi$$

$$p_0\, dY = \rho_0 v\, d\psi - p\, dx \qquad (45)$$

~~drag and lift functions. Now,~~ Assuming that the flow is irrotational, then by means of eqs. (11) and (12), it can be shown that the following relations hold:

$$p_0\, dX = (p+\rho u^2)dy - \rho uv\, dx = (p+\rho w^2)dy - \rho v\, d\phi$$

$$p_0\, dY = \rho uv\, dy - (p+\rho v^2)dx = \rho u\, d\phi - (p+\rho w^2)dx \qquad (46)$$

It is seen that by integrating eq. (46) along ~~at~~ any closed boundary, it ~~is~~ will give ~~[...]~~ the resultant of the pressure forces acting along the boundary and the rate of increase of momentum of the fluid passing out of the boundary. If there is a solid body in this boundary, then ~~it~~ this integral will give the lift + the drag acting on the body ~~by the fluid~~. Therefore, X + Y are sometimes called the drag + lift functions. From eq. (46) the following relations can be deduced:

$$p_0 (v\, dX - u\, dY) = p\, d\phi \qquad (47)$$

$$p_0 \frac{\rho}{\rho_0}(u\, dX + v\, dY) = (p+\rho w^2)\, d\psi \qquad (48)$$

Therefore, by writing $\frac{\partial\phi}{\partial X} = R$, + $\frac{\partial\phi}{\partial Y} = S$, eq. (47) gives:

$$R = \frac{\partial \phi}{\partial X} = \frac{p_0}{p} v \,, \qquad S = \frac{\partial \phi}{\partial Y} = -\frac{p_0}{p} u \tag{49}$$

~~Therefore,~~ The quantities R and S have the dimension of a velocity and R, S can be considered as components of a new velocity in the plane whose coordinates are denoted by X and Y. This relation between the xy plane and the XY plane is shown in Fig. 8. It is thus seen that if the undisturbed flow in the xy plane is in the positive x-direction, the undisturbed flow in the XY-plane will be in the negative Y-direction. Furthermore, if σ is defined as

$$\frac{\sigma}{\sigma_0} = \frac{p \frac{\rho}{\rho_0}}{p + \rho w^2} \tag{50}$$

eq. (48) gives ~~one has from eq. (48)~~,

$$\frac{\sigma}{\sigma_0} R = -\frac{\partial \psi}{\partial X} = -\frac{\frac{\rho}{\rho_0} u}{\frac{p + \rho w^2}{p_0}}$$

$$\frac{\sigma}{\sigma_0} S = \frac{\partial \psi}{\partial Y} = \frac{\frac{\rho}{\rho_0} v}{\frac{p + \rho w^2}{p_0}} \tag{51}$$

Comparing eq. (51) with eq. (12), it is evident that σ can be considered as the density of a fluid in the new XY plane.

Therefore there exists a complete reciprocity between the xy plane and the XY plane, as shown by H. Bateman (Ref. 6).

<u>Transformation starting with Incompressible Flow</u>

So far the relations obtained are general, i.e., they apply to fluids of arbitrary properties. However, since only the flow of incompressible fluid is well-known, it would be interesting to find the properties of the fluid in the X-Y plane if the fluid in the x-y plane is incompressible. If the fluid in the x-y plane is incompressible, then $\frac{\rho}{\rho_0} = 1$, and the Bernoulli theorem gives:

$$\frac{p + \frac{\rho}{2} w^2}{p_0} = 1 \tag{52}$$

In the X-Y plane, let P denote the pressure and Q^2 denote $R^2 + S^2$; then the generalized Bernoulli theorem, eq. (3), gives:

$$\int \frac{dP}{\sigma} + \frac{1}{2} Q^2 = \text{constant} \tag{53}$$

In view of eqs. (49), (50), and (52), eq. (53) can be written in the following form:

$$\frac{1}{\sigma_0} \int \frac{d\left(\frac{\sigma}{\sigma_0}\right)}{\left(\frac{\sigma}{\sigma_0}\right)} \frac{dP}{d\left(\frac{\sigma}{\sigma_0}\right)} + \frac{p_0}{\rho_0}\left(\frac{1}{4\frac{\sigma^2}{\sigma_0^2}} - \frac{1}{4}\right) = \text{constant} \tag{54}$$

By differentiating eq. (54) with respect to $\frac{\sigma}{\sigma_0}$, multiplying the resulting expression by $\frac{\sigma}{\sigma_0}$ and then integrating with respect to $\frac{\sigma}{\sigma_0}$, the following relation connecting the pressure P and the density σ for the fluid in the X-Y plane is obtained:

$$P = C - \frac{1}{2} \frac{p_0}{\rho_0} \sigma_0^2 \frac{1}{\sigma} \tag{55}$$

where C is the integration constant.

Comparing eq. (55) with the approximate adiabatic relation eq. (2), also noting eq. (46), it is evident that eqs. (55) and (2) are identical, if:

$$\frac{1}{2}\frac{p_0}{\rho_0} = A_0^2 = A_1^2\left[1-\left(\frac{Q_1}{A_1}\right)^2\right]$$

and $$C = P_1 + \frac{1}{2}\frac{p_0}{\rho_0}\sigma_0^2\frac{1}{\sigma_1} \qquad (56)$$

In eq. (56) A is the velocity of sound in the X-Y plane, and the subscript 1 refers to the conditions in the undisturbed flow. Hence, $\frac{Q_1}{A_1}$ is Mach's number of the undisturbed flow.

By using eqs. (52) and (49), the components of velocity in the X–Y plane can be expressed as:

$$\left.\begin{aligned} \frac{R}{Q_1} &= -\frac{v}{w_1}\;\frac{1-\frac{1}{2}\frac{\rho_0}{p_0}w_1^2}{1-\frac{1}{2}\frac{\rho_0}{p_0}w_1^2\left(\frac{w}{w_1}\right)^2} \\ \frac{S}{Q_1} &= \frac{u}{w_1}\;\frac{1-\frac{1}{2}\frac{\rho_0}{p_0}w_1^2}{1-\frac{1}{2}\frac{\rho_0}{p_0}w_1^2\left(\frac{w}{w_1}\right)^2} \end{aligned}\right\} \qquad (57)$$

Hence

$$\frac{Q}{Q_1} = \frac{w}{w_1}\;\frac{1-\frac{1}{2}\frac{\rho_0}{p_0}w_1^2}{1-\frac{1}{2}\frac{\rho_0}{p_0}w_1^2\left(\frac{w}{w_1}\right)^2} \qquad (58)$$

(*) [not new [illegible]] ~~In eqs. (57) and (58), it is assumed that the undisturbed flow in the XY plane is in the x direction and so the undisturbed flow in the XY plane is in the Y direction.~~ (*) The relation between w_1 & Q_1 can then be obtained from eqs. (56) & (57), that is:

$$\frac{1}{2}\frac{\rho_0}{p_0}w_1^2 = \frac{\left(\frac{Q_1}{A_1}\right)^2}{\left[1+\sqrt{1-\left(\frac{Q_1}{A_1}\right)^2}\right]^2} = \lambda \qquad (59)$$

Thus eq. (58) can be re-written as:

$$\frac{Q}{Q_1} = \frac{w}{w_1}\;\frac{1-\lambda}{1-\lambda\left(\frac{w}{w_1}\right)^2} \qquad (60)$$

Using eqs. (55), (56) & (58), the pressure coefficient Π in the X-Y plane which is defined by [illegible] $\Pi = \frac{P-P_1}{\frac{1}{2}\sigma_1 Q_1^2}$, can be expressed as

$$\Pi = \text{[illegible]} = \left(1+\frac{1}{2}\frac{\rho_0}{p_0}w_1^2\right)\frac{1-\left(\frac{w}{w_1}\right)^2}{1-\frac{1}{2}\frac{\rho_0}{p_0}w_1^2\left(\frac{w}{w_1}\right)^2}$$

~~[illegible] eq. (59)~~ Substituting the value of λ from eq (59) in the above expression, the following relation is obtained

$$\Pi = (1+\lambda)\,\frac{1-\left(\frac{w}{w_1}\right)^2}{1-\lambda\left(\frac{w}{w_1}\right)^2} \qquad (61)$$

must To find the coördinates X + Y in terms of x, y, eq. (46) be integrated. It is convenient in this case to use the complex potential of the incompressible flow in the xy plane. If

$$\phi + i\psi = w_1 G(x+iy) = w_1 G(z) \qquad (62)$$

then it can be shown with the aid of eq (52) that:

writing $\bar{Z}$ + $\bar{G}$ as the complex conjugate of Z and G,

$$\bar{Z} = X - iY = i\bar{z} - \frac{1}{2}\frac{\rho_0}{p_0} w_1^2 i \int \left(\frac{d\bar{G}}{d\bar{z}}\right)^2 d\bar{z}. \text{ Or}$$

$$e^{i\frac{\pi}{2}} \bar{Z} = z - \lambda \int \left(\frac{d\bar{G}}{d\bar{z}}\right)^2 d\bar{z} \qquad (63)$$

where the factor $e^{i\frac{\pi}{2}}$ will rotate the Z-plane through an angle equal to $\frac{\pi}{2}$ to make the directions of undisturbed flow in the Z-plane + in the z-plane coincide.

Comparing the set of eqs. (59), (60), (61) + (63) with the previous set of eqs. (33), (34), (35) + (32), it is evident that the two sets are identical except the change of notation. Therefore Bateman's transformation does not give any new results, but leads to the same expressions as those obtained by the hodograph method.

Concluding Remarks.

It is shown both in Section I + in Section II that starting from any solution of an incompressible fluid over a solid body, a solution of any nearly adiabatic flow over another approximately similar solid body can be calculated. The transformation from incompressible flow to compressible flow changes the shape of the body a small amount as represented by the correction terms in eqs. (32) + (63). Thus in order to investigate the effect of compressibility over the same body, it is necessary to use different functions $G(z)$ for different Mach's numbers as shown by the Busemann example given in Section I. This complicates the calculations to some extent, but the amount of labor involved, probably, is much less than the successive approximations devised by Janzen, Rayleigh, Poggi and Walther, especially at higher Mach's numbers.

The main difficulty of the method lies in its application to flow involving circulation, e.g. the flow over a lifting airfoil. Then if the ordinary complex potential function $G(z)$ for the incompressible fluid is used, the correction terms in eq. (32) & (33) are no longer uniform functions, that is, they do not return to their original value by increasing the argument of z by 2π. In other words, the boundary in compressible flow is no longer a closed curve. Therefore in order to study this type of problems, it is necessary to use a function $G(z)$ which does not give a closed boundary in the incompressible flow, but will give a closed boundary in the compressible flow when the correction term is added. The problem is thus more difficult, and requires further study.

The author expresses his gratitude to Dr. Th. von Kármán for suggesting the subject and for his kindly criticism during the course of the work.

References

*See next page!!!

1. Glauert, H.: "The Effect of Compressibility on the Lift of an Airfoil." Proc. Roy. Soc. (A) Vol. 118, p. 113, (1928), also R. & M. British A.R.C. 1135, (1928).

2. Poggi, L.: "Campo di velocità in una corrente piana di fluido compressibile." L'Aerotecnica, Vol. 12, pp. 1579-1593, (1932).

Part II. "Caso dei profile ottenutti con rappresentazione conforme dal cerchio ed in particolare dei profili Joukowski." L'Aerotecnica, Vol. 14, pp. 532-549, (1934).

3. Walther, P.A.: "Einfluss der Kompressibilität der Luft auf den Auftrieb eines Tragflügels." Trans. Central Aero-Hydro. Institute, No. 222, (1935).

"Evaluation of the Effect of Compressibility of Air on the Magnitude, on the Direction and on the Moment of the Lift of an Airplane Wing." Trans. Central Aero-Hydro. Institute. No. 274, (1936).

2 4. Kaplan, C.: "Two-dimensional Subsonic Compressible Flow past Elliptic Cylinders." N.A.C.A. T.R. No. 624, (1938).

5. Molenbroek, P.: "Über einige Bewegungen eines Gases bei Annahme eines Geschwindigkeits-potentials." Arch. d. Mathem. u. Phys. Grunert Hoppe (1890) Reihe Series 2, Bd. Vol. 9, p. 157. one word!!!

6. Tschapligin, S.A.: "Scientific Memoirs of the University of Moscow." (In Russian). (1902).

7. Clauser, F. and Clauser, M: "New Method of Solving the Equations for the Flow of a Compressible Fluid." Unpublished thesis at California Institute of Technology, (1937).

3 8. Demtchenko, B.: "Sur les mouvements lents des fluides compressibles." Comptes Rendus, Vol. 194, p. 1218, (1932).

"Variation de la résistance aux faibles vitesses sons l'influence de la compressibité." Comptes Rendus, Vol. 194, p. 1720, (1932).

References (Cont'd)

4 ~~9~~. Busemann, A.: "Die Expansionsberichtigung der kontraktionsziffer von Blenken." Forschung, Vol. ~~Bd.~~ 4, p. 186-187, (1933).

Hodographen methode der Gasdynamik, ZAMM, Vol. ~~Bd.~~ 12, p. 73-79, (1937).

5 ~~10~~. Hooker, S.G.: "The Two-Dimensional Flow of Compressible Fluids at Subsonic Speeds Past Elliptic Cylinders." R & M No. 1684, British A.R.C., (1936).

6 ~~11~~. Bateman, H.: "The Lift and Drag Functions for an Elastic Fluid in Two-Dimensional Irrotational Flow." Proc. National Acad. Sciences, Vol. 24, pp. 246-251, (1938).

* 1. ~~12~~. Glauert, H.: "The Effect of Compressibility on the Lift of an Airfoil." Proc. Roy. Soc. (A) Vol. 118, p. 113 (1928) also R. & M. British A.R.C. 1135 (1928).

Prandtl, L.: "Über Stromungen, deren Geschwindigkeiten mit der Schallgeschwindigkeit vergleichbar sind." Jour. of Aero. Research Institute, Tokyo Imp. Univ., No. 65, p. 14, (1930).

8 ~~13~~. Jacobs, E.N.: "Methods Employed in America for the Experimental Investigation of Aerodynamic Phenomena at High Speeds. Atti dei V. Convegni "Volta"; Le alte velocita in aivazione, p. 380, Reale Accademia D'Italia, Rome, (1936).

7 ~~12~~. Stack, J.: The Compressibility Burble NACA T.N. No. 543 (1935)

9 ~~14~~. Pinkerton, R.H.: Calculated + Measured Pressure distributions over the Midspan Section of the N.A.C.A. 4412 Airfoil N.A.C.A. T.R. No. 563 (1936)

Figure Legends

Fig. 1 Approximation to the adiabatic p-v relation by means of a tangent

Fig. 2 Relation between the maximum velocity (at which the pressure is zero) and Mach's number

Fig. 3. Variation of the parameter of transformation from incompressible flow to ~~compressible~~ flow with Mach's number

Fig. 4 Notations used in calculating the flow over an elliptical cylinder

Fig. 5 Flow over an elliptical cylinder with thickness ratio = 0.5 at Mach's number = 0.5
(a) Velocity distribution
(b) Pressure distribution

Fig. 6 Flow over an elliptical cylinder with thickness ratio = 0.1 at Mach's number = 0.857
(a) Velocity distribution
(b) Pressure distribution.

Fig. 7. Variation of critical Mach's number of an elliptical cylinder with thickness ratio.

Fig. 8 Relation of the velocity components in the x-y plane and X-Y plane.

TWO DIMENSIONAL SUBSONIC FLOW OF COMPRESSIBLE FLUIDS

by

Hsue-shen Tsien
California Institute of Technology

Introduction

Assuming that the pressure is a single-valued function of density only, the equations of two-dimensional irrotational motion of compressible fluids can be reduced to a single non-linear equation of the velocity potential. In the supersonic case, that is, in the case when the flow velocity is everywhere greater than that of local sound velocity, the problem is solved by Meyer & Prandtl and Busemann using the method of characteristics. The essential difficulty of this problem lies in the subsonic case, that is, in the case when the flow velocity is everywhere smaller than but near to the local sound velocity, because then the method of characteristics cannot be used. Glauert & Prandtl (Ref. 1) treated the case when the disturbance to the parallel rectilinear flow due to presence of a solid body is small. They were then able to linearize the differential equation for the velocity potential and obtained an equation very similar to those for the incompressible fluids. But there are usually stagnation points either in the surface of the body or in the field of flow, where the disturbance is no longer small. Hence, it is doubtful whether the linear theory can be applied to the flow near a stagnation point. On the same ground, the theory breaks down in case of bodies whose dimension across the stream is not small compared with the dimension parallel to the stream.

To treat cases in which the body is blunt nosed, Janzen and Rayleigh developed the method of successive approximations. This method was explained physically and put into a more convenient form by L. Poggi (Ref. 2) and P.A. Walther (Ref. 3). Recently C. Kaplan (Ref. 4) treated the case of flow over Joukowsky airfoils and elliptic cylinders, using Poggi's method. However, the method is rather tedious and the convergent very slow if the local velocity of sound is approached.

Molenbroek (Ref. 5) and Tschapligin (Ref. 6) suggested the use of the magnitude of velocity w and inclination β of velocity a chosen axis as independent variables, and were thus able to reduce the equation of velocity potential to a linear equation. This equation was solved by Tschapligin (Ref. 6) and recently put into a more convenient form by F. Clauser and M. Clauser (Ref. 7). The solution is essentially a series each term of which is a product of a hypergeometric function of w and a trigonometric function of β. The main difficulty in practical application of this solution is to obtain a proper set of boundary conditions in the plane of independent variables w, β and put the solution in a closed form.

Tschapligin (Ref. 6) shows that a great simplification of the equation in the hodograph plane results if the ratio of the specific heats of the gas is equal to -1. Since all real gases have their ratio of specific heats between 1 and 2, the value -1 seems without practical significance. It was Demtschenko (Ref. 8) and Busemann (Ref. 9) who made clear the meaning of this specific value of -1. They found that this really means to take the tangent of pressure-volume curve as an approximate to the curve itself. However, they limit themselves to use the tangent at the state of rest of the gas. Thus their theory can only be applied to flow with velocities up to one half that of sound velocity. Recently, during a personal discussion, Th. von Karman suggested to the author that the theory can be generalized to use the tangent at the state of gas corresponding to the undisturbed parallel flow. Thus the range of usefulness of the theory can be greatly extended. This is carried out in the first part of the present paper.

In the second part of this paper this theory based upon Demtschenko and Busemann's work is applied to the case of flow over elliptic cylinders and the results compared with those of S.G. Hooker (Ref.10) and C. Kaplan (Ref. 4). Furthermore, results calculated by Glauert's-Prandtl's linear theory are also included for comparison.

Recently, H. Bateman (Ref. 11) demonstrated a remarkable reciprocity of two fields of flow of two fluids related by a certain point transformation. It will be shown in the third part of this paper that the flow of incompressible fluid and the flow of compressible fluid approximated by the use of tangent to adiabatic pressure-volume curve can be interpreted as such a point transformation. It is thus possible to obtain a solution for compressible flow whenever a solution of incompressible flow is known. The difficulty, however, lies in the proper choice of this incompressible flow so that the solution of compressible flow will have a solid boundary desired. Fortunately, this transformed boundary is independent of the Mach's number of undisturbed flow. Therefore, by a modification of the procedure, two relations between the velocity and the pressure distributions over a body at low velocities and those at high velocities over the same body are obtained. This enables one to predict the characteristics of the flow over a body at any Mach's number whenever the low speed characteristics of the flow over the same body are known. The characteristics of the incompressible flow can either be obtained by the well-known method of conformal mapping or by experiments. Due to the fact that practical aerodynamic engineers usually have the low speed characteristics at hand and that high speed data have to be obtained by use of a costly high speed wind tunnel the above mentioned relations are believed to be of considerable use to them.

In the fourth part of the paper, the theory developed in Part III is applied to correlate airfoil data obtained by J. Stack (Ref. 12) in the N.A.C.A. 24" high speed wind tunnel. The agreement with theory is found to be satisfactory. Then this theory is applied to predict the compressibility effect on the lift and moment of N.A.C.A. 4412 airfoil using experimentally determined pressure distribution over the same airfoil at low speed. The result is again compared with the more simple Glauert-Prandtl theory.

I.

If p is the pressure, v is the specific volume and γ is the ratio of specific heats of a gas, the adiabatic relation $pv^{-\gamma}=$ constant is a curve in the p-v plane as shown in Fig. 1a. Now conditions near the point (p_1, v_1) which corresponds to the state of fluid at undisturbed flow can be approximated by the tangent to the curve at that point. The equation of the tangent at this point can be written as

$$p_1 - p = C(v_1 - v) = C(\rho_1^{-1} - \rho^{-1}) \tag{1}$$

where C is the slope of the tangent and ρ is the density of the gas. Now the slope C must be equal to the slope of the curve at the point $p_1 v_1$, therefore,

$$C = \left(\frac{dp}{dv}\right)_1 = \left(\frac{dp}{d\rho}\frac{d\rho}{dv}\right)_1 = -\left(\frac{dp}{d\rho}\right)_1 \rho_1^2 = -a_1^2\rho_1^2$$

where a_1 is the sound velocity ~~corresponds~~ corresponding to the conditions p_1, v_1 . Thus eq. (1) can be written as

$$p_1 - p = a_1^2\rho_1^2\left(\frac{1}{\rho} - \frac{1}{\rho_1}\right) \qquad (2)$$

This is an approximate pressure-density to adiabatic relation, and is shown in Fig. 1b with true adiabatic relation.

The Bernoullis' theorem for compressible fluids is

$$\frac{1}{2}w_2^2 - \frac{1}{2}w_3^2 = \int_2^3 \frac{dp}{\rho} \qquad (3)$$

where w is the velocity of the gas and the subscripts 2 and 3 denote two different states of the fluid. By substituting eq. (2) into eq. (3), the following relation is obtained:

$$\frac{1}{2}w_2^2 - \frac{1}{2}w_3^2 = \frac{1}{2}a_1^2\rho_1^2\left\{\frac{1}{\rho_3^2} - \frac{1}{\rho_2^2}\right\} \qquad (4)$$

Now if $w_3 = 0$, $w_2 = w$, $\rho_3 = \rho_0$, and $\rho_2 = \rho$, with the subscript 0 denotes the state of rest, eq. (4) gives:

$$\frac{a_1^2\rho_1^2}{\rho_0^2} + w^2 = \frac{\rho_1^2 a_1^2}{\rho^2} \qquad (5)$$

If the square of sound velocity a^2 is defined (as usually done) as the derivative of p with respect to ρ, eq. (2) gives:

$$a^2\rho^2 = \frac{dp}{d\rho}\rho^2 = a_1^2\rho_1^2 = \text{constant} \qquad (6)$$

Therefore, eq. (5) can be written as:

$$\left(\frac{\rho}{\rho_0}\right)^2 = 1 - \frac{w^2}{a^2} \qquad (7)$$

Similarly,

$$\left(\frac{\rho_0}{\rho}\right)^2 = 1 + \frac{w^2}{a_0^2} \qquad (8)$$

It is interesting to notice that from eq. (8) the density decreases as velocity increases, as expected. Thus eq. (6) shows that the local velocity of sound increases as the velocity increases. This is just opposite to the real gas, because in the case of an adiabatic flow of a real gas it is well known that the temperature of gas decreases as the velocity of gas is increased, and thus the local sound velocity also decreases. However, in the present approximate theory, the ratio $\frac{w}{a}$ or Mach's number still increases as the velocity increases, as can be seen by eq. (7). But this ratio only reaches the value unity when $\rho = 0$, or from eq. (8) when $w = \infty$. It is thus seen that the entire regime of flow is subsonic and thus the differential equation of velocity potential is always of elliptic type, that is, always of same type as the differential equation of velocity potential of incompressible fluids ~~of an incompressible~~. This is the reason why the complex representation of velocity potential and stream function is possible for all cases, as will be shown in following paragraphs. However, one should realize that the portion of tangent that could be used as an approximation to the true adiabatic relation is that portion which lies in the first quadrant. Thus the upper limit of velocity for practical application of the theory is when $p = 0$. By using eqs. (17) and (18), this upper limit is found to be

$$\left(\frac{w}{w_1}\right)_{max} = \frac{1}{\left(\frac{w_1}{a_1}\right)}\sqrt{\left(\frac{p_1}{a_1^2 \rho_1}\right)^2 - \left\{1-\left(\frac{w_1}{a_1}\right)^2\right\}} \tag{9}$$

Since the point (p_1, ρ_1) lies on the true adiabatic curve, the relation $a_1^2 = \gamma \frac{p_1}{\rho_1}$ can be used, and eq. (9) becomes:

$$\left(\frac{w}{w_1}\right)_{max} = \frac{1}{\left(\frac{w_1}{a_1}\right)}\sqrt{\left(\frac{1}{\gamma}+1\right)^2 - \left\{1-\left(\frac{w_1}{a_1}\right)^2\right\}} \tag{10}$$

This relation is plotted in Fig. 2. Since for most practical cases it is not likely that the ratio $\left(\frac{w}{w_1}\right)$ will rise to values much higher than 2 , p will remain positive, and this theory will give an approximate solution.

If the flow is irrotational, there exists a velocity potential ϕ such that

$$\frac{\partial \phi}{\partial x} = u, \quad \frac{\partial \phi}{\partial y} = v \tag{11}$$

where u, v are the components of w in x and y direction, respectively. The equation of continuity,

$$\frac{\partial}{\partial x}\left(\frac{\rho}{\rho_0} u\right) + \frac{\partial}{\partial y}\left(\frac{\rho}{\rho_0} v\right) = 0$$

will be satisfied, if the stream function ψ is introduced such that

$$\frac{\rho}{\rho_0} u = \frac{\partial \psi}{\partial y}, \quad -\frac{\rho}{\rho_0} v = \frac{\partial \psi}{\partial x} \tag{12}$$

Now if the angle of inclination of the velocity w to the x axis is β, eqs. (11) and (12) give:

$$d\phi = w\cos\beta\, dx + w\sin\beta\, dy$$
$$d\psi = -w\frac{\rho}{\rho_0}\sin\beta\, dx + w\frac{\rho}{\rho_0}\cos\beta\, dy \qquad (13)$$

Solving for dx and dy.

$$dx = \frac{\cos\beta}{w}d\phi - \frac{\sin\beta}{w}\frac{\rho_0}{\rho}d\psi$$
$$dy = \frac{\sin\beta}{w}d\phi + \frac{\cos\beta}{w}\frac{\rho_0}{\rho}d\psi \qquad (14)$$

So long as the correspondence between the physical plane and hodograph plane is one to one, or mathematically $\frac{\partial(x,y)}{\partial(u,v)} \neq 0$, one can express x and y as functions of w, β and so ϕ and ψ as functions of w, β. Thus,

$$d\phi = \phi'_w\, dw + \phi'_\beta\, d\beta \qquad (15)$$
$$d\psi = \psi'_w\, dw + \psi'_\beta\, d\beta$$

where primes indicate the derivative with respect to variables indicated as subscripts. Now substituting eq. (15) into eq. (14), one has:

$$dx = \left(\frac{\cos\beta}{w}\phi'_w - \frac{\sin\beta}{w}\frac{\rho_0}{\rho}\psi'_w\right)dw + \left(\frac{\cos\beta}{w}\phi'_\beta - \frac{\sin\beta}{w}\frac{\rho_0}{\rho}\psi'_\beta\right)d\beta$$

$$dy = \left(\frac{\sin\beta}{w}\phi'_w + \frac{\cos\beta}{w}\frac{\rho_0}{\rho}\psi'_w\right)dw + \left(\frac{\sin\beta}{w}\phi'_\beta + \frac{\cos\beta}{w}\frac{\rho_0}{\rho}\psi'_\beta\right)d\beta \qquad (16)$$

Since the left-hand side of eqs.(16) are exact differentials, one can apply the reciprocity relation, and obtain:

$$\frac{\partial}{\partial\beta}\left(\frac{\cos\beta}{w}\phi'_w - \frac{\sin\beta}{w}\frac{\rho_0}{\rho}\psi'_w\right) = \frac{\partial}{\partial w}\left(\frac{\cos\beta}{w}\phi'_\beta - \frac{\sin\beta}{w}\frac{\rho_0}{\rho}\psi'_\beta\right)$$
$$\frac{\partial}{\partial\beta}\left(\frac{\sin\beta}{w}\phi'_w + \frac{\cos\beta}{w}\frac{\rho_0}{\rho}\psi'_w\right) = \frac{\partial}{\partial w}\left(\frac{\sin\beta}{w}\phi'_\beta + \frac{\cos\beta}{w}\frac{\rho_0}{\rho}\psi'_\beta\right) \qquad (17)$$

Carrying out these differentiations and simplifying with the aid of eq. (7), eq. (17) gives:

$$-\frac{\sin\beta}{w}\phi'_w-\frac{\cos\beta}{w}\frac{\rho_0}{\rho}\psi'_w=-\frac{\cos\beta}{w^2}\phi'_\beta+\frac{\sin\beta}{w^2}\frac{\rho}{\rho_0}\psi'_\beta$$
$$\frac{\cos\beta}{w}\phi'_w-\frac{\sin\beta}{w}\frac{\rho_0}{\rho}\psi'_w=-\frac{\sin\beta}{w^2}\phi'_\beta-\frac{\cos\beta}{w^2}\frac{\rho}{\rho_0}\psi'_\beta \tag{18}$$

Solving for ϕ'_w and ψ'_β,

$$\phi'_w=-\frac{\rho}{\rho_0}\frac{1}{w}\psi'_\beta$$
$$\phi'_\beta=\frac{\rho_0}{\rho}w\,\psi'_w \tag{19}$$

Eq. (19) can be further simplified by introducing a new variable ω, such that

$$d\omega=\frac{\rho}{\rho_0}\frac{dw}{w} \tag{20}$$

Then eq. (19) becomes:

$$\phi'_\omega=-\psi'_\beta$$
$$\phi'_\beta=\psi'_\omega \tag{21}$$

This can be easily recognized as the Riemann-Cauchy's differential equations and thus $\phi+i\psi$ must be an analytic function of $\omega-i\beta$. However, for convenience of calculation, another new set of independent variables $U=W\cos\beta$, $V=W\sin\beta$ are introduced where $W=a_0e^{\omega}$.

Then eq. (21) can be written as:

$$\frac{\partial\phi}{\partial U}=\frac{\partial\psi}{\partial(-V)}$$
$$\frac{\partial\phi}{\partial(-V)}=-\frac{\partial\psi}{\partial U} \tag{22}$$

and also by integrating eq. (20),

$$W=\frac{2a_0w}{\sqrt{a_0^2+w^2}+a_0} \tag{23}$$

and

$$w = \frac{4a_0^2 W}{4a_0^2 - W^2} \tag{24}$$

Substituting into eq. (8), we have:

$$\frac{\rho_0}{\rho} = \frac{4a_0^2 + W^2}{4a_0^2 - W^2} \tag{25}$$

Eqs.(22), (23), (24) and (25) are the basic equations of the present theory. One recognizes eqs.(22) as the Riemann-Cauchy differential equations, and thus the complex potential $F = \phi + i\psi$ must be an analytic function of $\overline{W} = U - iV$, or

$$\begin{aligned} \phi + i\psi &= F(U - iV) = F(\overline{W}) \\ \phi - i\psi &= \bar{F}(U + iV) = \bar{F}(W) \end{aligned} \tag{26}$$

Now, it is necessary to find the values of x and y corresponds to a given set of values of U and V, i.e., to find the transformation from hodograph plane to physical plane. By using eqs. (24) and (25), eq. (14) can be written as:

$$\begin{aligned} dx &= \frac{U\cdot d\phi}{W^2}\left\{1 - \frac{W^2}{4a_0^2}\right\} - \frac{V\cdot d\psi}{W^2}\left\{1 + \frac{W^2}{4a_0^2}\right\} \\ dy &= \frac{V\cdot d\phi}{W^2}\left\{1 - \frac{W^2}{4a_0^2}\right\} + \frac{U\cdot d\psi}{W^2}\left\{1 + \frac{W^2}{4a_0^2}\right\} \end{aligned} \tag{27}$$

where $W^2 = U^2 + V^2$. These equations can be combined into one equation by means of eq. (26). Thus,

$$dz = dx + i\,dy = \frac{dF}{\overline{W}} - \frac{W\cdot d\bar{F}}{4a_0^2} \tag{28}$$

Hence, if an analytic function $F(\overline{W})$ is given for each value of W, the corresponding real velocity w can be calculated by eq. (24). Then the coordinate of the point in the physical plane at which this velocity occurs can be calculated by means of integrating eq. (28). The pressure at this point is then given by eq. (2). However, there is an objection to this procedure. That is, the investigator has no way of telling whether with the chosen function $F(\bar{W})$, he is going to obtain the desired shape of solid boundary and flow pattern which he is interested in. In other words, this procedure still suffers the difficulty of boundary conditions as it is common to all hodograph methods.

However, due to the particular simple relation of eq. (28), one can ascertain approximately the resulting shape of the body by starting with the particular function,

$$F(\bar{W}) = \phi + i\psi = W_1 \, G(\zeta) \tag{29}$$

where W_1 is the transformed undisturbed velocity to be interpreted by eq. (23), and ζ is the complex coordinate $\xi + i\eta$. This function is so chosen as to give the flow of incompressible fluid over the desired body shape in coordinates ξ and η. Now interpret the real velocity in $\xi-\eta$ plane of the incompressible fluid as the transformed velocity W in the hodograph plane for compressible fluid. Then it is known that

$$\bar{W} = W_1 \frac{dG(\zeta)}{d\zeta} \tag{30}$$

Thus

$$W = W_1 \frac{d\bar{G}(\bar{\zeta})}{d\bar{\zeta}} \tag{31}$$

Substituting eqs. (30) and (31) into eq. (28), one has:

$$dz = d\zeta - \frac{1}{4}\left(\frac{W_1}{a_0}\right)^2\left[\frac{d\bar{G}(\bar{\zeta})}{d\bar{\zeta}}\right]^2 d\bar{\zeta}$$

Integrating,

$$z = \zeta - \frac{1}{4}\left(\frac{W_1}{a_0}\right)^2 \int \left(\frac{d\bar{G}}{d\bar{\zeta}}\right)^2 d\bar{\zeta} \tag{32}$$

Therefore, the complex coordinate in the physical plane of compressible fluid is equal to the corresponding complex coordinate in the physical plane of incompressible fluid plus a correction term. Since this correction term is usually small, the resulting shape of the body will be quite similar to the one in incompressible fluid. The factor in the correction term depends upon the Mach's number of the undisturbed flow only. This can be seen by means of eqs. (7), (8), and (23), because from those equations one has the following relation:

$$\frac{1}{4}\left(\frac{W_1}{a_0}\right)^2 = \frac{\left(\frac{w_1}{a_1}\right)^2}{\left\{1+\sqrt{1-\left(\frac{w_1}{a_1}\right)^2}\right\}} \tag{33}$$

where $\left(\frac{w_1}{a_1}\right)$ is the Mach's number of the undisturbed flow.

To calculate the velocity in physical plane, $\bar{W}$ is first obtained from eq. (30) and then with eq. (23):

$$\frac{w}{w_1} = \left\{1-\frac{1}{4}\left(\frac{W_1}{a_0}\right)^2\right\}\frac{\frac{|W|}{W_1}}{1-\frac{1}{4}\left(\frac{W_1}{a_0}\right)^2\left\{\frac{|W|}{W_1}\right\}^2} \tag{34}$$

Then the pressure can be calculated by using eq. (2). With some calculation the following expression for pressure at any point is obtained:

$$\varpi = \frac{p-p_0}{\frac{1}{2}\rho_1 w_1^2} = \frac{2}{\left(\frac{w_1}{a_1}\right)^2}\left\{1-\sqrt{1+\left(\frac{w_1}{a_1}\right)^2\left\{\left(\frac{w}{w_1}\right)^2-1\right\}}\right\} \tag{35}$$

II.

In this part, the theory developed in Part I will be applied to calculate the flow over an elliptic cylinder at zero angle of attack. The incompressible flow over an elliptic cylinder in the complex coordinate ζ can be obtained by applying the Joukowsky's transformation to the flow over a circular cylinder in the complex coordinate ζ' with its center located at the origin. Therefore, the function $F(\bar{W})$ or $W_1 G(\zeta)$ can be written as:

$$F = W_1 \left[\zeta' + \frac{a^2}{\zeta'}\right] \qquad (36)$$
$$\bar{F} = W_1 \left[\bar{\zeta}' + \frac{a^2}{\bar{\zeta}'}\right]$$

and
$$\zeta = \zeta' + \frac{1}{\zeta'} \qquad (37)$$

It is convenient to carry out the computation by using the ζ' coordinates. Thus eq. (32) is rewritten in the following form:

$$z = \left(\zeta' + \frac{1}{\zeta'}\right) - \frac{1}{4}\left(\frac{W_1}{a_0}\right)^2 \int \left(\frac{dG}{d\bar{\zeta}'}\right)^2 \frac{d\bar{\zeta}'}{\frac{d\bar{\zeta}}{d\bar{\zeta}'}} \qquad (38)$$

If one limits one's self to calculate the conditions over the surface of the elliptic cylinder only,

$$\zeta' = b e^{i\theta} \qquad (39)$$
$$\bar{\zeta}' = b e^{-i\theta}$$

where θ is the argument as shown in Fig. 3 and b is the radius of the circular section in the ζ'-plane which determines the thickness ratio of the elliptic section in the ζ-plane. Substituting eqs. (36), (37) and (39) into eq. (38), and carrying out the integration, the following expressions for the x and y coordinates corresponding to ζ' are obtained by separating the real and imaginary parts,

$$x=\left(b+\frac{1}{b}\right)\cos\theta-\frac{W_1^2}{4a_0^2}\left\{b(1+b^2)\cos\theta+\frac{(b^2-1)^2}{4}\log\frac{(b^2-1)^2+4b^2\sin^2\theta}{(b^2+2b\cos\theta+1)}\right\}$$

$$y=\left(b-\frac{1}{b}\right)\sin\theta+\frac{W_1^2}{4a_0^2}\left\{b(1-b^2)\sin\theta+\frac{(b^2-1)^2}{2}\tan^{-1}\frac{2b\sin\theta}{b^2-1}\right\} \qquad (40)$$

The horizontal and vertical semi-axis of the near elliptic section can then be calculated by substituting $\theta=0$ and $\theta=\frac{\pi}{2}$ respectively into eq. (40). The thickness ratio δ is thus obtained as:

$$\delta=\left(\frac{b^2-1}{b^2+1}\right)\left\{\frac{1+\frac{W_1^2}{4a_0^2}\left[-b^2+\frac{b(b^2-1)}{2}\tan^{-1}\frac{2b}{b^2-1}\right]}{1-\frac{W_1^2}{4a_0^2}\left[b^2+\frac{b(b^2-1)}{2}\left(\frac{b^2-1}{b^2+1}\right)\log\left(\frac{b-1}{b+1}\right)\right]}\right\} \qquad (41)$$

For a given thickness ratio and Mach's number of undisturbed stream, the value of $\frac{1}{4}\left(\frac{W_1}{a_0}\right)^2$ is first computed by means of eq. (33), and then eq. (41) is solved graphically for b

After b is obtained, the coordinate x and y for each value of θ can be computed by using eq. (40). It is fortunate that the values of x, y so obtained lies very close to the true elliptic section as shown in Fig. 4 Hence, the velocity and the pressure distribution obtained by using eqs. (34) and (35) are considered as those over the true elliptic sections.

Calculations for two thickness ratios, $\delta = 0.5$ and $\delta = 0.1$, are carried out and the results shown in Fig. 5, Fig. 6, Fig. 7 and Fig. 8, together with those of Kaplan (Ref. 4) and Hooker (Ref. 10) included. In order to compare the results with that obtained from less exact theory of Glauert and Prandtl, these computations are made (Ref. 1), and the results also included in Fig. 5 to Fig. 8. It is seen from those figures that the effect of compressibility of fluids shown by the present theory is more pronounced than those shown by Kaplan's and Hooker's calculation. In view of the fact that Kaplan and Hooker carried their calculation only to first approximation, and higher approximations will increase the value somewhat, it is probable that the values given by the present theory are nearer to the true value. As to the results given by Prandtl-Glauert theory, it is seen that the theory breaks down near the stagnation point and the results are poorer for thick sections than those for thin sections. However, in case of thin sections this very approximate theory does give satisfactory information for a point not too close to the stagnation point.

If the velocity of flow over a body is gradually increased, the maximum local velocity in the field will also be increased. When a certain velocity of the undisturbed stream is reached the maximum local velocity reaches the local velocity of sound. Then shock waves usually appear and the drag of the body suddenly increases. Therefore, this velocity is of considerable interest to practical engineers and is usually referred to as the critical velocity of the body. It is shown by Kaplan (Ref. 4) and others that at this critical condition the ratio of maximum velocity w_{max} in the field to the undisturbed velocity w_1, is related to the Mach's number $\frac{w_1}{a_1}$ of undisturbed stream in the following manner:

$$\left(\frac{w_{max}}{w_1}\right) = \left\{\frac{2}{\gamma+1}\frac{1}{\left(\frac{w_1}{a_1}\right)^2} + \frac{\gamma-1}{\gamma+1}\right\}^{\frac{1}{2}} \tag{42}$$

w_{max} in the flow over an elliptic cylinder at zero angle of attack occurs at the top of the cylinder. Using eqs. (34) and (33) the value of $\frac{w_{max}}{w_1}$ is found to be:

$$\frac{w}{w_1} = \frac{\left(\frac{2b^2}{b^2+1}\right)^2}{1 + \frac{\left[1-\left(\frac{2b^2}{b^2+1}\right)^2\right]\left(\frac{w_1}{a_1}\right)^2}{2\left(1+\sqrt{1-\left(\frac{w_1}{a_1}\right)^2}\right)\sqrt{1-\left(\frac{w_1}{a_1}\right)^2}}} \tag{43}$$

Equating eqs. (42) and (43), the equation for calculating the critical Mach's number $\left(\frac{w_1}{a_1}\right)$ of the undisturbed stream for each value of b is:

$$\frac{2}{\gamma+1}\frac{1}{\left(\frac{w_1}{a_1}\right)_{cri}} + \frac{\gamma-1}{\gamma+1} = \frac{\left(\frac{2b^2}{b^2+1}\right)^2}{1 + \frac{\left[1-\left(\frac{2b^2}{b^2+1}\right)^2\right]\left(\frac{w_1}{a_1}\right)^2}{2\left(1+\sqrt{1-\left(\frac{w_1}{a_1}\right)^2}\right)\sqrt{1-\left(\frac{w_1}{a_1}\right)^2}}} \tag{44}$$

Knowing $\left(\frac{w_1}{a_1}\right)_{cri}$ for each value of b, the corresponding value of δ can be calculated by means of eqs. (34) and (41). Fig. 9 shows the result of this calculation with Kaplan's

$$\mathcal{R} = \frac{\partial \phi}{\partial X} = \frac{p_0}{p_0} v, \quad \mathcal{S} = \frac{\partial \phi}{\partial Y} = -\frac{p_0}{p} u \tag{49}$$

Therefore, the quantities $\mathcal{R}$ and $\mathcal{S}$ have the dimension of a velocity and so $\mathcal{R}$ and $\mathcal{S}$ can be considered as a new velocity in the plane whose coordinates are denoted by X and Y. This relation between the xy plane and the X ~~and~~ Y plane is shown in Fig. 10. It is thus seen that if the undisturbed stream in the xy plane is in the positive x- direction, the undisturbed stream in the XY- plane will be in the negative Y- direction. Furthermore, if

$$\frac{\sigma}{\sigma_0} = \frac{p \frac{\rho}{\rho_0}}{p + \rho w^2} \tag{50}$$

one has from eq. (48),

$$\begin{aligned} \frac{\sigma}{\sigma_0} \mathcal{R} &= -\frac{\partial \psi}{\partial X} = -\frac{\frac{\rho}{\rho_0} u}{\frac{p + \rho w^2}{p_0}} \\ \frac{\sigma}{\sigma_0} \mathcal{S} &= \frac{\partial \psi}{\partial Y} = \frac{\frac{\rho}{\rho_0} v}{\frac{p + \rho w^2}{p_0}} \end{aligned} \tag{51}$$

Then there exists a complete reciprocity between the xy plane and the XY plane, as shown by H. Bateman (Ref. 11).

Comparing eq. (51) with eq. (12), it is evident that σ can be considered as the density of a fluid in the new XY plane.

So far the relations obtained are general, i.e., they apply to fluids of arbitrary properties. However, since only the flow of incompressible fluid has been studied exhaustively, it would be interesting to find out what will be the flow in the XY plane if the flow in the xy plane is incompressible. If the fluid in the xy plane is incompressible, then $\frac{\rho}{\rho_0} = 1$, and the Bernoulli's head is:

$$H = \frac{p + \frac{\rho}{2} w^2}{p_0} = 1 \tag{52}$$

Now in the XY plane, let P denote the pressure and Q^2 denote $R^2 + S^2$; then Bernoulli's theorem gives:

$$\int \frac{dP}{\sigma} + \frac{1}{2} Q^2 = \text{constant} \tag{53}$$

In view of eqs. (49), (50), and (52), eq. (53) can be written in the following form:

$$\frac{1}{\sigma_0} \int \frac{d\left(\frac{\sigma}{\sigma_0}\right)}{\left(\frac{\sigma}{\sigma_0}\right)} \frac{dP}{d\left(\frac{\sigma}{\sigma_0}\right)} + \frac{p_0}{\rho_0}\left(\frac{1}{4 \frac{\sigma^2}{\sigma_0^2}} - \frac{1}{4}\right) = \text{constant} \tag{54}$$

By differentiating eq. (54) with respect to $\frac{\sigma}{\sigma_0}$, multiplying the resulting expression by $\frac{\sigma}{\sigma_0}$ and then integrating with respect to $\frac{\sigma}{\sigma_0}$, one has the following relation connecting the pressure P and density σ for the fluid in the XY plane:

$$P = \text{Constant} - \frac{1}{2} \frac{p_0}{\rho_0} \sigma_0^2 \frac{1}{\sigma} \tag{55}$$

Comparing eq. (55) with the approximate adiabatic relation eq. (2), also noting eq. (16), it is evident that eqs. (55) and (2) are identical if

~~(56)~~

$$\frac{1}{2}\frac{p_0}{\rho_0} = A_0^2 = A_1^2\left[1-\left(\frac{Q_1}{A_1}\right)^2\right] \tag{56}$$

and Constant = $P_1 + \frac{1}{2}\frac{p_0}{\rho_0}\sigma_0^2\frac{1}{\sigma_1}$

In eq. (56) A is the velocity of sound in the XY plane, and the subscript 1 refers to the conditions in the undisturbed parallel stream. Hence, $\frac{Q_1}{A_1}$ is the Mach's number of the undisturbed parallel stream.

To find the coordinates X and Y in terms of x, y, eq. (46) has to be integrated. It is convenient in this case to use the complex potential of incompressible flow in the xy plane. Thus, if

$$\phi + i\psi = w_1 G(x+iy) = w_1 G(z) \tag{57}$$

Then it can be easily shown with the aid of eq. (52) that

$$\bar{Z} = X - iY = \frac{1}{2}\frac{\rho_0}{p_0} w_1^2\, i\int\left(\frac{dG}{dz}\right)^2 dz \tag{58}$$

It is interesting to notice that eq. (58) is identical to the correction terms in eq. (32) except a factor. However, there is a fundamental difference between the method explained in Part I and the method explained in this Part. The previous method will give different solid boundaries for different Mach's number of the undisturbed flow even if the same function G is used. Eq. (58) shows that if the same function G is used, the resulting solid boundaries are all geometrically similar.

By using eqs. (52) and (49), the components of velocity in the XY plane can be expressed as:

$$\left.\begin{aligned}\frac{R}{Q_1} &= -\frac{v}{w_1}\,\frac{1-\frac{1}{2}\frac{\rho_0}{p_0}w_1^2}{1-\frac{1}{2}\frac{\rho_0}{p_0}w_1^2\left(\frac{w}{w_1}\right)^2}\\ \frac{S}{Q_1} &= \frac{u}{w_1}\,\frac{1-\frac{1}{2}\frac{\rho_0}{p_0}w_1^2}{1-\frac{1}{2}\frac{\rho_0}{p_0}w_1^2\left(\frac{w}{w_1}\right)^2}\end{aligned}\right\} \tag{59}$$

Or

$$\frac{Q}{Q_1} = \frac{w}{w_1}\,\frac{1-\frac{1}{2}\frac{\rho_0}{p_0}w_1^2}{1-\frac{1}{2}\frac{\rho_0}{p_0}w_1^2\left(\frac{w}{w_1}\right)^2} \tag{60}$$

In eqs. (59) and (60), it is assumed that the undisturbed flow in the xy plane is in the x-direction and so the undisturbed flow in the XY plane is in the Y direction. The relation between w_1 and Q_1 can then be obtained from eqs. (56) and (57), that is:

$$\frac{1}{2}\frac{\rho_0}{p_0}w_1^2 = \frac{\left[1+\sqrt{1-\left(\frac{Q_1}{A_1}\right)^2}\right]^2}{\left(\frac{Q_1}{A_1}\right)^2} = f\left(\frac{Q_1}{A_1}\right) \tag{61}$$

Thus, eq. (60) can be rewritten as:

$$\frac{Q}{Q_1} = \frac{w}{w_1}\,\frac{1-f}{1-f\cdot\left(\frac{w}{w_1}\right)^2} \tag{62}$$

For a given condition of flow in the XY plane, eqs. (56) and (61) will determine the characteristics of the flow in the xy plane. Using eqs. (55), (56) and (50), the pressure in the XY plane can be evaluated as:

$$\Pi = \frac{P - P_1}{\frac{1}{2}\rho_1 Q_1^2} = \left(1 + \frac{1}{2}\frac{\rho_0}{p_0} w_1^2\right) \frac{1 - \left(\frac{w}{w_1}\right)^2}{1 - \frac{1}{2}\frac{\rho_0}{p_0} w_1^2 \left(\frac{w}{w_1}\right)^2}$$

Thus by means of eq. (61),

$$\Pi = (1 + f) \frac{1 - \left(\frac{w}{w_1}\right)^2}{1 - f\left(\frac{w}{w_1}\right)^2} \qquad (63)$$

Thus as the method is developed one starts with a certain function $G(z)$ representing the incompressible flow in the xy plane, we first calculate the coordinates $X\,Y$ and thus the solid boundary in the $X\,Y$ plane. Then with the desired value of $\left(\frac{Q_1}{A_1}\right)$ for flow in the $X\,Y$ plane, the velocity and pressure distribution can be calculated by means of eqs. (61), (62) and (63). Thus the flow problem of approximately adiabatic compressible fluid is solved by means of an auxiliary function $G(z)$ for flow of incompressible fluids. However, this procedure suffers the defect of difficulty in obtaining the desired solid boundary compared with the first method developed in Part I, because now Z is not related to z by a small correction as is in eq. (32), but by a transformation given by eq. (68).

However, a much more useful form of eqs. (62) and (63) can be obtained if one remembers that the solid boundary of the flow in the $X\,Y$ plane is independent of Mach's number, but only on $G(z)$, or in other words, on $\frac{w}{w_1}$. Now substitute eq. (61)

into eq. (62) and let $\left(\frac{Q_1}{A_1}\right) \to 0$. Then

$$\left(\frac{Q}{Q_1}\right)_0 = \frac{1}{\left(\frac{w}{w_1}\right)}$$

or

$$\frac{w}{w_1} = \frac{1}{\left(\frac{Q}{Q_1}\right)_0} \tag{64}$$

where $\left(\frac{Q}{Q_1}\right)_0$ denotes the velocity for flow in the XY plane when $\left(\frac{Q_1}{A_1}\right) = 0$. Then substitute eq. (64) into eq. (62). The following relation between the velocity of incompressible flow and the velocity of compressible flow over the same solid boundary is obtained:

$$\left(\frac{Q}{Q_1}\right) = \left(\frac{Q}{Q_1}\right)_0 \frac{1-f}{\left(\frac{Q}{Q_1}\right)_0^2 - f} \tag{65}$$

where f is a function of $\left(\frac{Q_1}{A_1}\right)$ given by eq. (61). Similarly eq. (63) can be written as:

$$\Pi = (1+f)\frac{\left(\frac{Q}{Q_1}\right)_0^2 - 1}{\left(\frac{Q}{Q_1}\right)_0^2 - f} \tag{66}$$

If Π for $\left(\frac{Q_1}{A_1}\right) = 0$ is denoted by Π_0 , then it is well known that

$$\Pi_0 = 1 - \left(\frac{Q}{Q_1}\right)_0^2$$

or

$$\left(\frac{Q}{Q_1}\right)_0^2 = 1 - \Pi_0 \tag{67}$$

Substituting eq. (67) into eq. (66), we have:

$$\Pi = \frac{(1+f)\Pi_0}{f - (1-\Pi_0)} \tag{68}$$

Eqs. (65) and (68) together with eq. (61) thus enable one to calculate the velocity and pressure distribution over a body in compressible fluid once the velocity and pressure distribution over the same body in incompressible fluid is known. The latter can be obtained by either well-known analytical means or by experiments.

If eq. (61) is substituted into eq. (68) and only first order terms of $\left(\frac{Q_1}{A_1}\right)$ is retained, then

$$\Pi \cong \frac{\Pi_0}{\sqrt{1-\left(\frac{Q_1}{A_1}\right)^2}} \tag{69}$$

Thus, if C_L and C_{L_0} are the lift coefficients of an airfoil at $\left(\frac{Q_1}{A_1}\right)$ and at $\left(\frac{Q_1}{A_1}\right)=0$ and the same angle of attack respectively, one has from eq. (69):

$$C_L = \frac{C_{L_0}}{\sqrt{1-\left(\frac{Q_1}{A_1}\right)^2}} \tag{70}$$

which checks with the Prandtl-Glauert theory (Ref. 1).

IV.

By inverting eq. (68), one has:

$$\Pi_0 = \frac{(f-1)\,\Pi}{(f+1)-\Pi} \tag{71}$$

If the theory developed in Part III is true, then the value of Π_0 calculated from Π at different values of Mach's number $\left(\frac{Q_1}{A_1}\right)$ of undisturbed flow should fall on the same curve. This is done for the case of upper surface of NACA 4412 airfoil tested by J. Stack (Ref. 2) at $-0°/5'$ angle of attack. The Π are replotted in Fig. 10, while the Π_0 curves calculated from Π curves are shown in Fig. 11. The derived values of Π_0 lie fairly close to one curve. Since the data for this calculation is taken from a small scale figure, the deviations must be largely due to inaccuracies involved. Fig. 11 can thus be taken as an experimental check of the theory.

One of the immediate possible applications of the theory developed in Part III is to determine the critical speed of a body. If $\left(\frac{Q}{Q_1}\right)_{0\,max}$ and $\Pi_{0\,min}$ are the maximum speed and minimum pressure over a body. Then by eqs. (65), (67) and (42), the equation for critical Mach's number is obtained as:

$$\left(\frac{Q_1}{A_1}\right)^2_{cr} = \frac{2}{\gamma+1}\,\frac{1}{\left(\frac{Q}{Q_1}\right)^2_{0\,max}\left\{\frac{f-1}{f-\left(\frac{Q}{Q_1}\right)^2_{0\,max}}\right\}^{\frac{2}{\gamma+1}}} \qquad (72)$$

or

$$\left(\frac{Q_1}{A_1}\right)^2_{cr} = \frac{2}{\gamma+1}\,\frac{1}{\left(1-\Pi_{0\,min}\right)\left\{\frac{f-1}{(f-1)+\Pi_{0\,min}}\right\}^{\frac{2}{\gamma+1}}} \qquad (73)$$

In both eqs. (72) and (73), f is a function of $\left(\frac{Q_1}{A_1}\right)$ given by eq. (61). Eq. (73) is plotted in Fig. 12, together with a curve given by E. N. Jacobs (Ref. 13) based upon the Glauert-Prandtl theory. It is thus seen that Jacobs's curve gives a critical Mach's number, which is a little too high. This difference is, of course, to be expected, due to the higher order approximation of the present theory.

References

1. Glauert, H. : "The Effect of Compressibility on the Lift of an Airfoil." Proc. Roy. Soc. (A) Vol. 118, p. 113, (1928), also R. & M. British A.R.C. 1135, (1928).

2. Poggi, L. : "Campo di velocita in una corrente peana di fluido compressibile." L'Aerotecnica, Vol. 12, pp. 1579-1593, (1932).

   Part II. "Caso dei profile ottenutti con rappresentazious conforme dal cerchio id in particolare dei profili." Joukowski - L'Aerotecnica, Vol. 14, pp. 532-549, (1934).

3. Walther, P.A. : "Einfluss der Kompressibilität der Luft auf den Auftrieb eines Tragflügels." Trans. Central Aero-Hydro. Institute, No. 222, (1935).

   "Evaluation of the Effect of Compressibility of Air on the Magnitude, on the Direction and on the Moment of the Lift of an Airplane Wing." Trans. Central Aero-Hydro. Institute. No. 274, (1936).

4. Kaplan, C. : "Two-dimensional Subsonic Compressible Flow past Elliptic Cylinders." N.A.C.A. T.R. No. 624, (1938).

5. Molenbroek, P. : "Über einige Bewegungen eines Gases bei Annahme eines Geschwindigkeits potentials." Arch. d. Mathem. u. Phys. Grunert Hoppe (1890) Reihe 2, Bd. 9, p. 157.

6. Tschapligin, A. : "Scientific Memoirs of the University of Moscow." (In Russian). (1902).

7. Clauser, F. and Clauser, M: "New Method of Solving the Equations for the Flow of a Compressible Fluid." Unpublished thesis at California Institute of Technology, (1937).

8. Demtchenko, B. : "Sur les mouvements lents des fluides compressibles." Comptes Rendus, Vol. 194, p. 1218, (1932).

   "Variation de la resistance aux faibles vitesses sons l'influence de la compressibite." Comptes Rendus, Vol. 194, p. 1720, (1932).

References (Cont'd)

9. Busemann, A.: "Die Expansionsberichtigung der kontraktionsziffer von Blenken." Forschung, Bd. 4, p. 186-187, (1933).

Hodographen methode der Gasdynamik, ZAMM, Bd. 12, p. 73-79, (1937).

10. Hooker, S.G.: "The Two-Dimensional Flow of Compressible Fluids at Subsonic Speeds Past Elliptic Cylinders." R & M No. 1684, British A.R.C., (1936).

11. Bateman, H.: "The Lift and Drag Functions for an Elastic Fluid in Two-Dimensional Irrotational Flow." Proc. National Acad. Sciences, Vol. 24, pp. 246-251, (1938).

1. ~~12.~~ Glauert, H.: "The Effect of Compressibility on the Lift of an Airfoil." Proc. Roy. Soc. (A) Vol. 118, p. 113 (1928) also R. & M. British A.R.C. 1135 (1928).

Prandtl, L.: "Über Stromungen, deren Geschwindigkeiten mit der Schallgeschwindigkeit vergleichbar sind." Jour. of Aero. Research Institute, Tokyo Imp. Univ., No. 65, p. 14, (1930).

13. Jacobs, E.N.: "Methods Employed in America for the Experimental Investigation of Aerodynamic Phenomena at High Speeds. Atti dei V. Convegni "Volta"; Le alte velocita in aivazione, p. 380, Reale Accademia D'Italia, Rome, (1936).

12. Stack, J.: The Compressibility Burble NACA T.N. 543 (1935)

FIGURE LEGENDS

Fig. 1. Approximation to the adiabatic relation by means of a tangent.

Fig. 2. Relation between the maximum velocity (at which the pressure is zero) and Mach's number.

Fig. 3. Variation of the parameter of transformation from incompressible flow to compressible flow with Mach's number.

Fig. 4. Notations used in calculating the flow over an elliptical cylinder.

Fig. 5. Flow over an elliptical cylinder with thickness ratio = 0.5 at Mach's number = 0.5.

(a) Velocity distribution

(b) Pressure distribution

Fig. 6. Flow over an elliptical cylinder with thickness ratio = 0.1 at Mach's number = 0.857.

(a) Velocity distribution

(b) Pressure distribution

Fig. 7. Variation of critical Mach's number of an elliptical cylinder with thickness ratio.

Fig. 8. Relation of the velocity components in the plane and plane.

Section 2

Similarity Laws of Hypersonic Flows

SIMILARITY LAWS OF HYPERSONIC FLOWS 35

by Hsue-Shen Tsien 78

Introduction

Hypersonic flows are flow fields where the fluid velocity is much larger than the velocity of propagation of small disturbances, the velocity of sound. Th. von Kármán (Ref. 1) has pointed out that the dynamics of hypersonic flows is similar to Newton's corpuscular theory of aerodynamics in many ways. The pressure acting on an inclined surface is thus greater than the free stream pressure by a quantity which is approximately proportional to the square of the angle of inclination instead of the usual linear law for conventional supersonic flows. E. Sänger (Ref. 2) has, in fact, used this concept to design the optimum wing and body shapes for hypersonic flight at extreme speeds.

Recently von Kármán (Ref. 3) has obtained the similarity laws for transonic flows where the fluid velocity is very near to the velocity of sound. He deduced these laws by using an affine transformation of the fluid field so that the differential equations of the flows are reduced to a single non-dimensional equation. In this paper, the same method is used to derive the similarity laws for hypersonic flows. These laws will be, perhaps, useful in correlating the experimental data to be obtained in the near future by hypersonic wind tunnels now under construction.

Differential Equation for Hypersonic Flows

If u, v are the components of velocity in the x, y directions and a is the local velocity of sound, the differential equations for irrotational two-dimensional motion are

$$\left(1-\frac{u^2}{a^2}\right)\frac{\partial u}{\partial x} - \frac{uv}{a^2}\left(\frac{\partial u}{\partial y}+\frac{\partial v}{\partial x}\right) + \left(1-\frac{v^2}{a^2}\right)\frac{\partial v}{\partial y} = 0 \tag{1}$$

$$\frac{\partial v}{\partial x} - \frac{\partial u}{\partial y} = 0 \tag{2}$$

Now if a slender body is present in an otherwise uniform stream of velocity U in the x-direction, equations (2) is satisfied by introducing the disturbance velocity potential φ defined as

$$u = U + \frac{\partial \varphi}{\partial x}$$
$$v = \frac{\partial \varphi}{\partial y} \tag{3}$$

If a_0 is the velocity of sound for the gas at rest and a^0 is the velocity of sound corresponding to the free stream velocity U, then there are the following relations:

$$a^2 = a_0^2 - \frac{\gamma-1}{2}(u^2+v^2) = a_0^2 - \frac{\gamma-1}{2}\left[U^2 + 2U\frac{\partial \varphi}{\partial x} + \left(\frac{\partial \varphi}{\partial x}\right)^2 + \left(\frac{\partial \varphi}{\partial y}\right)^2\right]$$
$$a^{0^2} = a_0^2 - \frac{\gamma-1}{2}U^2 \tag{4}$$

where γ is the ratio of the specific heats.

For hypersonic flows over a slender body, both a^0 and $\frac{\partial \varphi}{\partial x}, \frac{\partial \varphi}{\partial y}$ are small in comparison with U. By substituting equations (3) and (4) into equation (1) and retaining terms up to second order, one has

$$\left[1-(\gamma+1)M^0\frac{1}{a^0}\frac{\partial \varphi}{\partial x} - \frac{\gamma-1}{2}\frac{1}{a^{0^2}}\left(\frac{\partial \varphi}{\partial y}\right)^2 - M^{0^2}\right]\frac{\partial^2 \varphi}{\partial x^2} - 2M^0\frac{1}{a^0}\frac{\partial \varphi}{\partial y}\frac{\partial^2 \varphi}{\partial x \partial y}$$
$$+\left[1-(\gamma-1)M^0\frac{1}{a^0}\frac{\partial \varphi}{\partial x} - \frac{\gamma+1}{2}\frac{1}{a^{0^2}}\left(\frac{\partial \varphi}{\partial y}\right)^2\right]\frac{\partial^2 \varphi}{\partial y^2} = 0. \tag{5}$$

Here M^0 is the Mach number of the free stream or

$$M^0 = \frac{U}{a^0} \tag{6}$$

Similarity Laws in Two-Dimensional Flow

If $2b$ is the length or chord of the body and δ the thickness of the body, the non-dimensional coördinates ξ and η can be defined as

$$x = b\xi \qquad y = b\left(\frac{\delta}{b}\right)^n \eta \tag{7}$$

where n is the exponent yet to be determined. von Kármán (Ref. 1) has shown that for hypersonic flow over a slender body the variation of fluid velocity due to the presence of the body is limited within a narrow region close to the body, the hypersonic boundary layer. Therefore, in order to investigate this velocity variation, one must expand the coördinate normal to the ~~the body~~ surface ~~of~~. This is similar to the case of ordinary viscous boundary layer, where Prandtl's simplified boundary layer equation is obtained from the exact Navier-Stokes equations by a coördinate expansion normal to the surface of the body. From this reasoning then, n must be positive so that η is much greater than (θ/δ). This surmise is substantiated by the later calculations to be shown presently.

By 75767

The appropriate non-dimensional form for the velocity potential φ is

$$\varphi = a^0 b \frac{1}{M^0} f(\xi, \eta) \tag{8}$$

By substituting equations (7) and (8) into equation (5), one has

$$\left(\frac{\delta}{b}\right)^{2n}\left[1-(\gamma+1)\frac{\partial f}{\partial \xi} - \frac{\gamma-1}{2}\frac{1}{M^{02}\left(\frac{\delta}{b}\right)^{2n}}\left(\frac{\partial f}{\partial \eta}\right)^2\right]\frac{\partial^2 f}{\partial \xi^2} - M^{02}\left(\frac{\delta}{b}\right)^{2n}\frac{\partial^2 f}{\partial \xi^2} - 2\frac{\partial f}{\partial \eta}\frac{\partial^2 f}{\partial \xi \partial \eta}$$

$$+\left[1-(\gamma-1)\frac{\partial f}{\partial \xi} - \frac{\gamma+1}{2}\frac{1}{M^{02}\left(\frac{\delta}{b}\right)^{2n}}\left(\frac{\partial f}{\partial \eta}\right)^2\right]\frac{\partial^2 f}{\partial \eta^2} = 0 \tag{9}$$

The boundary conditions at infinity requires that the flow velocity to be $U_\circ$. Thus

$$\frac{\partial f}{\partial \xi} = \frac{\partial f}{\partial \eta} = 0 \qquad \text{at } \infty \tag{10}$$

If the slender body is a ~~symmetrical~~ one, then the condition at the surface of the body can be written as

$$\left(\frac{\partial \varphi}{\partial y}\right)_{y=0} = a^\circ M^\circ \left(\frac{\delta}{b}\right) h(\xi) \qquad \text{for } -1 < \xi < 1 \tag{11}$$

(for $-1<\xi<1$)

where $h(\xi)$ is a given function describing the thickness distribution along the length of the body. Equation (11) can be converted into the following form by means of equations (7) and (8),

$$\left(\frac{\partial f}{\partial \eta}\right)_{\eta=0} = M^{\circ 2} \left(\frac{\delta}{b}\right)^{1+n} h(\xi) \tag{12}$$

Since the body is thin, $\frac{\delta}{b}$ is very small. Therefore the first group of terms in equation (9) is negligible in comparison with the rest. Then both the differential equation and the boundary conditions can be made to contain only a single parameter if one sets

$$n = 1 \tag{13}$$

That is, if

$$M^\circ \frac{\delta}{b} = K$$

then equation (9) becomes

$$\left[1-(\gamma-1)\frac{\partial f}{\partial \xi} - \frac{\gamma+1}{2}\frac{1}{K^2}\left(\frac{\partial f}{\partial \eta}\right)^2\right]\frac{\partial^2 f}{\partial \eta^2} = K^2\frac{\partial^2 f}{\partial \xi^2} + 2\frac{\partial f}{\partial \eta}\frac{\partial^2 f}{\partial \xi \partial \eta} \tag{14}$$

and the boundary conditions become

$$\frac{\partial f}{\partial \xi} = \frac{\partial f}{\partial \eta} = 0 \qquad \text{at } \infty$$

and

$$\left(\frac{\partial f}{\partial \eta}\right)_{\eta=0} = K^2 h(\xi), \qquad -1<\xi<1 \tag{15}$$

The meaning of this similarity law is the following: If a series of bodies having the same thickness distribution but different thickness ratios (δ/b) are put into flows of different Mach numbers M^0 such that the products of M^0 and (δ/b) remain constant and equal to K, then the flow patterns are similar in the sense that they are governed by the same function $f(\xi, \eta)$, determined by equations (14) and (15).

If p_0 is the stagnation pressure, p^0 the free stream pressure, and p the local pressure, then

$$p^0 = p_0\left[1+\frac{\gamma-1}{2}M^{0^2}\right]^{-\frac{\gamma}{\gamma-1}}$$

$$p = p_0\left[1+\frac{\gamma-1}{2}\frac{u^2+v^2}{a^2}\right]^{-\frac{\gamma}{\gamma-1}}$$

In notations introduced previously, and by retaining terms of proper magnitudes, one can write the expression for the local pressure as

$$p = p^0\left[1-(\gamma-1)\frac{\partial f}{\partial \xi}-\frac{\gamma-1}{2}\frac{1}{K^2}\left(\frac{\partial f}{\partial \eta}\right)^2\right]^{\frac{\gamma}{\gamma-1}} \quad (16)$$

The drag D of the body can then be calculated. It is given by the following expression:

$$D = 2\int_{-b}^{b}(p)_{\eta=0}\,h(\xi)\left(\frac{\delta}{b}\right)dx = 2b\,p^0\left(\frac{\delta}{b}\right)\int_{-1}^{1}\left[1-(\gamma-1)\frac{\partial f}{\partial \xi}-\frac{\gamma-1}{2}\frac{1}{K^2}\left(\frac{\partial f}{\partial \eta}\right)^2\right]_{\eta=0}^{\frac{\gamma}{\gamma-1}} h(\xi)\,d\xi$$

If one wishes to compute the drag coefficient C_D, then one uses

$$C_D = \frac{D}{\frac{\rho^0}{2}U^2(2b)} = \frac{1}{M^{0^3}}\left\{\frac{2}{\gamma}K\int_{-1}^{1}\left[1-(\gamma-1)\frac{\partial f}{\partial \xi}-\frac{\gamma-1}{2}\frac{1}{K^2}\left(\frac{\partial f}{\partial \eta}\right)^2\right]_{\eta=0}^{\frac{\gamma}{\gamma-1}} h(\xi)\,d\xi\right\} \quad (17)$$

For a given thickness distribution, the quantity within the brackets is only a function of K, the similarity parameter. Therefore one can write

$$C_D = \frac{1}{M^{\circ 3}} \Delta(K) = \frac{1}{M^{\circ 3}} \Delta\left(M^\circ \frac{f}{b}\right) \qquad (18)$$

Similarly, one obtains for the lift coefficient C_L of the lift L the following law:

$$C_L = \frac{L}{\frac{\rho^\circ}{2} U^{\circ}(2b)} = \frac{1}{M^{\circ 2}} \Lambda(K) = \frac{1}{M^{\circ 2}} \Lambda\left(M^\circ \frac{f}{b}\right) \qquad (19)$$

These similarity laws show that for bodies of the same thickness distributions at angles of attack proportional to the thickness ratio (f/b), the quantities $(C_D M^{\circ 3})$ and $(C_L M^{\circ 2})$ are single functions of the parameter $K = M^\circ \frac{f}{b}$.

Equations (18) and (19) agree with the results of the more limited linearized theory of Ackeret (Ref. 4). According to this theory, for similar bodies in the sense stated above the drag coefficient and the lift coefficient are given by

$$C_D \sim \frac{\left(\frac{f}{b}\right)^2}{\sqrt{M^{\circ 2} - 1}}$$

$$C_L \sim \frac{\left(\frac{f}{b}\right)}{\sqrt{M^{\circ 2} - 1}}$$

For hypersonic flows of very large values of M°, these expressions reduces to

$$C_D \sim \left(\frac{f}{b}\right)^2 / M^\circ \qquad (20)$$

$$C_L \sim \left(\frac{f}{b}\right) / M^\circ \qquad (21)$$

Equations (20) and (21) agrees with equations (18) and (19). Equations (18) and (19) are, however, more general and complete.

Axially Symmetrical Flows

For axially symmetrical flows, the ordinate y is the radial distance from the axis to the point concerned. Then a similar analysis leads to the following differential equation and boundary conditions:

$$\left[1-(\gamma-1)\frac{\partial f}{\partial \xi}-\frac{\gamma+1}{2}\frac{1}{K^2}\left(\frac{\partial f}{\partial \eta}\right)^2\right]\frac{\partial^2 f}{\partial \eta^2}+\left[1-(\gamma-1)\frac{\partial f}{\partial \xi}-\frac{\gamma-1}{2}\frac{1}{K^2}\left(\frac{\partial f}{\partial \eta}\right)^2\right]\frac{1}{\eta}\frac{\partial f}{\partial \eta}$$

$$=2\frac{\partial f}{\partial \eta}\frac{\partial^2 f}{\partial \xi \partial \eta}+K^2\frac{\partial^2 f}{\partial \xi^2} \qquad (22)$$

$$\frac{\partial f}{\partial \xi}=\frac{\partial f}{\partial \eta}=0 \quad \text{at } \infty \qquad (23)$$

$$\left(\eta\frac{\partial f}{\partial \eta}\right)_{\eta=0}=K^2 h(\xi) \quad , \text{for } -1<\xi<1 \qquad (24)$$

where $h(\xi)$ is the distribution function for cross-sectional areas along the length of the body and $K=M^0\left(\frac{t}{b}\right)$ as for two-dimensional flows.

The drag coefficient C_D refered to the maximum cross section of the body is then governed by the following similarity law:

$$C_D=\frac{1}{M^{02}}\Delta\left(M^0\frac{t}{b}\right) \qquad (25)$$

California Institute of Technology

References

1) Th. von Kármán, "The Problem of Resistance in Compressible Fluids" Proceedings of the 5th Volta Congress, pp. 275-277, Rome (1936)

2) E. Sänger, "Gleitkörper für sehr hohe Fluggeschwindigkeiten" German Patent 411/42, Berlin, (1937)

3) Th. von Kármán, "Similarity Laws of Transonic Flows" To be published soon.

4) See for instance Durand "Aerodynamic Theory" pp. 234-236, Vol. III, J. Springer, Berlin (1935).

SIMILARITY LAWS OF HYPERSONIC FLOWS

by Hsue-shen Tsien

Introduction

Hypersonic flows are flow fields where the fluid velocity is much larger than the velocity of propagation of small disturbances, the velocity of sound. Th. von Kármán (Ref. 1) has pointed out that in many ways the dynamics of hypersonic flows is similar to Newton's corpuscular theory of aerodynamics. The pressure acting on an inclined surface is thus greater than the free stream pressure by a quantity which is approximately proportional to the square of the angle of inclination instead of the usual linear law for conventional supersonic flows. E. Sänger (Ref. 2) has, in fact, used this concept to design the optimum wing and body shapes for hypersonic flight at extreme speeds.

Recently, von Kármán (Ref. 3) has obtained the similarity laws for transonic flows where the fluid velocity is very near to the velocity of sound. He deduced these laws by using an affine transformation of the fluid field so that the differential equations of the flows are reduced to a single non-dimensional equation. In this paper, the same method is used to derive the similarity laws for hypersonic flows. These laws will be, perhaps, useful in correlating the experimental data to be obtained in the near future by hypersonic wind tunnels now under construction.

Differential Equation for Hypersonic Flows

If u, v are the components of velocity in the x, y directions and a is the local velocity of sound, the differential equations for irrotational two-dimensional motion are

$$\left(1-\frac{u^2}{a^2}\right)\frac{\partial u}{\partial x} - \frac{uv}{a^2}\left(\frac{\partial u}{\partial y}+\frac{\partial v}{\partial x}\right) + \left(1-\frac{v^2}{a^2}\right)\frac{\partial v}{\partial y} = 0 \qquad (1)$$

$$\frac{\partial v}{\partial x} - \frac{\partial u}{\partial y} = 0$$

Now if a slender body is present in an otherwise uniform stream of velocity V in the x-direction, equations (2) is satisfied by introducing the disturbance velocity potential φ defined as

$$u = V + \frac{\partial \varphi}{\partial x}, \qquad v = \frac{\partial \varphi}{\partial y} \qquad (3)$$

If a_o is the velocity of sound for the gas at rest and a^o is the velocity of sound corresponding to the free stream velocity V, then there are the following relations:

$$a^2 = a_o^2 - \frac{\gamma-1}{2}(u^2+v^2) = a_o^2 - \frac{\gamma-1}{2}\left[V^2 + 2V\frac{\partial \varphi}{\partial x} + \left(\frac{\partial \varphi}{\partial x}\right)^2 + \left(\frac{\partial \varphi}{\partial y}\right)^2\right] \qquad (4)$$

$$a^{o2} = a_o^2 - \frac{\gamma-1}{2}V^2$$

where γ is the ratio of the specific heats.

For hypersonic flows over a slender body, both a^o and $\frac{\partial \varphi}{\partial x}, \frac{\partial \varphi}{\partial y}$ are small in comparison with V. By substituting equations (3) and (4) into equation (1) and retaining terms up to second order, one has

$$\left[1-(\gamma+1)M^o\frac{1}{a^o}\frac{\partial \varphi}{\partial x} - \frac{\gamma-1}{2}\frac{1}{a^{o2}}\left(\frac{\partial \varphi}{\partial y}\right)^2 - M^{o2}\right]\frac{\partial^2 \varphi}{\partial x^2} - 2M^o\frac{1}{a^o}\frac{\partial \varphi}{\partial y}\frac{\partial^2 \varphi}{\partial x \partial y}$$

$$+\left[1-(\gamma-1)M^o\frac{1}{a^o}\frac{\partial \varphi}{\partial x} - \frac{\gamma+1}{2}\frac{1}{a^{o2}}\left(\frac{\partial \varphi}{\partial y}\right)^2\right]\frac{\partial^2 \varphi}{\partial y^2} = 0 \qquad (5)$$

Here M^o is the Mach number of the free stream or

$$M^o = \frac{V}{a^o} \qquad (6)$$

Similarity Laws in Two-Dimensional Flow

If 2b is the length or chord of the body and δ the thickness of the body, the non-dimensional coordinates ξ and η can be defined as

$$
\begin{aligned}
x &= b\xi \\
y &= b\left(\frac{\delta}{b}\right)^n \eta
\end{aligned}
\tag{7}
$$

where n is the exponent yet to be determined. von Kármán (Ref. 1) has shown that for hypersonic flow over a slender body the variation of fluid velocity due to the presence of the body is limited within a narrow region close to the body, the hypersonic boundary layer. Therefore, in order to investigate this velocity variation, one must expand the coordinate normal to the surface of the body. This is similar to the case of ordinary viscous boundary layer, where Prandtl's simplified boundary layer equation is obtained from the exact Navier-Stokes equations by a coordinate expansion normal to the surface of the body. From this reasoning then, n must be positive so that η is much greater than (y)/(b). This surmise is substantiated by the later calculations to be shown presently.

The appropriate non-dimensional form for the velocity potential φ is

$$
\varphi = a^0 b \frac{1}{M^0} f(\xi, \eta) \tag{8}
$$

By substituting equations (7) and (8) into equation (5), one has

$$
\begin{aligned}
&\left(\frac{\delta}{b}\right)^{2n}\left[1-(\gamma+1)\frac{\partial f}{\partial \xi}-\frac{\gamma-1}{2}\frac{1}{M^{0\,2}\left(\frac{\delta}{b}\right)^{2n}}\left(\frac{\partial f}{\partial \eta}\right)^2\right]\frac{\partial^2 f}{\partial \xi^2}-M^{0\,2}\left(\frac{\delta}{b}\right)^{2n}\frac{\partial^2 f}{\partial \xi^2}-2\frac{\partial f}{\partial \eta}\frac{\partial^2 f}{\partial \xi \partial \eta} \\
&+\left[1-(\gamma-1)\frac{\partial f}{\partial \xi}-\frac{\gamma+1}{2}\frac{1}{M^{0\,2}\left(\frac{\delta}{b}\right)^{2n}}\left(\frac{\partial f}{\partial \eta}\right)^2\right]\frac{\partial^2 f}{\partial \eta^2}=0
\end{aligned}
\tag{9}
$$

The boundary conditions at infinity require that the flow velocity be V. Thus

$$\frac{\partial f}{\partial \xi} = \frac{\partial f}{\partial \eta} = 0 \qquad at \qquad \infty \tag{10}$$

If the slender body is a symmetrical one, then the condition at the surface of the body can be written as

$$\left(\frac{\partial \varphi}{\partial y}\right)_{y=0} = a^{0} M^{0} \left(\frac{\delta}{b}\right) h(\xi) \tag{11}$$

where h(ξ) for $-1 < \xi < 1$ is a given function describing the thickness distribution along the length of the body. Equation (11) can be converted into the following form by means of equations (7) and (8).

$$\left(\frac{\partial f}{\partial \eta}\right)_{\eta=0} = M^{0\,2} \left(\frac{\delta}{b}\right)^{1+n} h(\xi) \tag{12}$$

Since the body is thin, $\left(\frac{\delta}{b}\right)$ is very small. Therefore the first group of terms in equation (9) is negligible in comparison with the rest. Then both the differential equation and the boundary conditions can be made to contain only a single parameter if one sets

$$n = 1 \tag{13}$$

That is, if

$$M^{0} \frac{\delta}{b} = K$$

then equation (9) becomes

$$\left[1 - (\gamma-1)\frac{\partial f}{\partial \xi} - \frac{\gamma+1}{2}\frac{1}{K^2}\left(\frac{\partial f}{\partial \eta}\right)^2\right]\frac{\partial^2 f}{\partial \eta^2} = K^2 \frac{\partial^2 f}{\partial \xi^2} + 2\frac{\partial f}{\partial \eta}\frac{\partial^2 f}{\partial \xi \partial \eta} \tag{14}$$

and the boundary conditions become

$$\frac{\partial f}{\partial \xi} = \frac{\partial f}{\partial \eta} = 0 \qquad at \qquad \infty \tag{15}$$

and

$$\left(\frac{\partial f}{\partial \eta}\right)_{\eta=0} = K^2 h(\xi) \quad , \text{ for } \quad -1 < \xi < 1$$

The meaning of this similarity law is the following: If a series of bodies having the same thickness distribution but different thickness ratios (δ/b) are put into flows of different Mach numbers M^o such that the products of M^o and (δ/b) remain constant and equal to K, then the flow patterns are similar in the sense that they are governed by the same function $f(\xi, \eta)$, determined by equations (14) and (15).

If p_o is the stagnation pressure, p^o the free stream pressure, and p the local pressure, then

$$p^o = p_o \left[1 + \frac{\gamma-1}{2} M^{o2}\right]^{-\frac{\gamma}{\gamma-1}}$$

$$p = p_o \left[1 + \frac{\gamma-1}{2} \frac{u^2+v^2}{a^2}\right]^{-\frac{\gamma}{\gamma-1}}$$

In notations introduced previously and by retaining terms of proper magnitudes, one can write the expression for the local pressure as

$$p = p^o \left[1 - (\gamma-1)\frac{\partial f}{\partial \xi} - \frac{\gamma-1}{2}\frac{1}{K^2}\left(\frac{\partial f}{\partial \eta}\right)^2\right]^{\frac{\gamma}{\gamma-1}} \tag{16}$$

The drag D of the body can then be calculated. It is given by the following expression:

$$D = 2\int_{-b}^{b} (p)_{\eta=0}\, h(\xi)\left(\frac{\delta}{b}\right) dx = 2b\, p^o \left(\frac{\delta}{b}\right) \int_{-1}^{1} \left[1 - (\gamma-1)\frac{\partial f}{\partial \xi} - \frac{\gamma-1}{2}\frac{1}{K^2}\left(\frac{\partial f}{\partial \eta}\right)^2\right]_{\eta=0}^{\frac{\gamma}{\gamma-1}} h(\xi)\, d\xi$$

If one wishes to compute the drag coefficient C_D, then one uses

$$C_D = \frac{D}{\frac{\rho^o}{2} V^2 (2b)} = \frac{1}{M^{o3}} \left\{ \frac{2}{\gamma} K \int_{-1}^{1} \left[1 - (\gamma-1)\frac{\partial f}{\partial \xi} - \frac{\gamma-1}{2}\frac{1}{K^2}\left(\frac{\partial f}{\partial \eta}\right)^2\right]_{\eta=0}^{\frac{\gamma}{\gamma-1}} h(\xi)\, d\xi \right\} \tag{17}$$

For a given thickness distribution, the quantity within the brackets is only a function of K, the similarity parameter. Therefore, one can write

$$C_D = \frac{1}{M^{o3}} \Delta(K) = \frac{1}{M^{o3}} \Delta\left(M^o \frac{\delta}{b}\right) \tag{18}$$

Similarly, one obtains for the lift coefficient C_L of the lift L the following law:

$$C_L = \frac{L}{\frac{\rho^o}{2} V^2 (2b)} = \frac{1}{M^{o2}} \Lambda(K) = \frac{1}{M^{o2}} \Lambda\left(M^o \frac{\delta}{b}\right) \tag{19}$$

These similarity laws show that for bodies of the same thickness distribution at angles of attack proportional to the thickness ratio (δ/b), the quantities ($C_D M^{o3}$) and $C_L M^{o2}$) are functions of the single parameter $K = M^o \frac{\delta}{b}$.

Equations (18) and (19) agree with the results of the more limited linearized theory of Ackeret (Ref. 4). According to this theory, for similar bodies in the sense stated above the drag coefficient and the lift coefficient are given by

$$C_D \sim \frac{\left(\frac{\delta}{b}\right)^2}{\sqrt{M^{o2}-1}}$$

$$C_L \sim \frac{\frac{\delta}{b}}{\sqrt{M^{o2}-1}}$$

For hypersonic flows of very large values of M^o, these expressions reduce to

$$C_D \sim (\delta/b)^2 / M^o \tag{20}$$

$$C_L \sim (\delta/b) / M^o \tag{21}$$

Equations (20) and (21) agree with equations (18) and (19). Equations (18) and (19) are, however, more general and complete.

Axially Symmetrical Flows

For axially symmetrical flows, the ordinate y is the radial distance from the axis to the point concerned. Then a similar analysis leads to the following differential equation and boundary conditions:

$$\left[1-(\gamma-1)\frac{\partial f}{\partial \xi} - \frac{\gamma+1}{2}\frac{1}{K^2}\left(\frac{\partial f}{\partial \eta}\right)^2\right]\frac{\partial^2 f}{\partial \eta^2} + \left[1-(\gamma-1)\frac{\partial f}{\partial \xi} - \frac{\gamma-1}{2}\frac{1}{K^2}\left(\frac{\partial f}{\partial \eta}\right)^2\right]\frac{1}{\eta}\frac{\partial f}{\partial \eta} = 2\frac{\partial f}{\partial \eta}\frac{\partial^2 f}{\partial \xi \partial \eta} + K^2\frac{\partial^2 f}{\partial \xi^2} \qquad (22)$$

$$\frac{\partial f}{\partial \xi} = \frac{\partial f}{\partial \eta} = 0 \quad \text{at} \quad \infty \qquad (23)$$

$$\left(\eta\frac{\partial f}{\partial \eta}\right)_{\eta=0} = K^2 h(\xi), \quad \text{for} \quad -1<\xi<1 \qquad (24)$$

where $h(\xi)$ is the distribution function for cross-sectional areas along the length of the body and $K = M^0 \frac{\delta}{b}$ as for two-dimensional flow.

The drag coefficient C_D referred to the maximum cross section of the body is then governed by the following similarity law:

$$C_D = \frac{1}{M^{0\,2}}\,\Delta\left(M^0\frac{\delta}{b}\right) \qquad (25)$$

References

1. Th. von Kármán - "The Problem of Resistance in Compressible Fluids" Proceedings of the 5th Volta Congress, pp. 275-277, Rome (1936).
2. E. Sänger, "Gleitkörper für sehr hohe Fluggeschwindigkeiten" German Patent 411/42, Berlin, (1939).
3. Th. von Kármán, "Similarity Laws of Transonic Flows" , to be published soon.
4. See for instance Durand "Aerodynamic Theory" pp. 234-236, Vol. III, J. Springer, Berlin (1935).

July 8, 1946

Prof. E. Reissner
Department of Mathematics
Massachusetts Institute of Technology
Cambridge, Mass.

Dear Prof. Reissner:

Enclosed is the manuscript of the short note on hypersonic flows which we talked about during my recent visit to M.I.T. I am submitting it for consideration of publication in your JMP. I hope it is not too late to make the deadline.

Our discussion a few days ago was most enjoyable and I am looking forward to many fruitful collaborations after this Fall. Best regards to Mrs. Reissner and Johnny, I am

Sincerely yours,

H. S. Tsien

HST:mo
Encl.

Section 3

The "Limiting Line" in Mixed Subsonic and Supersonic Flows of Compressible Fluids

The "Limiting Line" in mixed Subsonic and Supersonic Flows of Compressible Fluids

It is well-known that the vorticity for any fluid element is constant if the fluid is non-viscous and the change of states of the fluid is isentropic. When a solid body is placed in a uniform stream, the flow well ahead of the body is irrotational. Then if the flow is further assumed to be isentropic, the vorticity will remain to be zero over the whole field of flow. In other words, the flow is irrotational. For such flow over a solid body of finite dimensions, it is shown by T. Theodorsen (Ref. 1) that the solid body experiences no resistance. If the fluid has a small viscosity, its effect will be limited in the boundary layer over the solid body and the body will have a drag due to the skin friction. This type of essentially irrotational isentropic flow is generally observed for a stream-lined body placed in a uniform stream, if the velocity of the stream is kept below the so-called "critical speed." At the "critical speed" or rather at a certain value of the ratio of the velocity of undisturbed stream and the corresponding velocity of sound, shock waves appear. (This phenomenon is called the "compressibility burble".) Along a shock wave, the change of states of the fluid is no more isentropic, although still adiabatic. This results in an increase in entropy of the fluid and generally introduces vorticity in an originally irrotational flow. The increase in entropy of the fluid is, of course, the consequence of changing part of the mechanical energy into heat energy. In other words, the part of fluid affected by the shock wave has much less mechanical energy. Therefore with the appearance of shock waves, the wake of the stream line body is very much widened, (Ref. 2, 3) and the drag increases drastically. Furthermore the accompanied change in the pressure distribution over the body changes the moment acting on it and in case of an airfoil decreases the lift force.

All these consequences of the break down of isentropic irrotational flow are generally undesirable in applied aerodynamics. Its occurrence should be delayed as much as possible by modifying the shape or contour of the body. However, such endeavor will be very much facilitated if the cause or criterion for the break down could be found.

Criterion for the Break Down of Isentropic Irrotational Flow

G. I. Taylor and C.F. Sharman (Ref. 2) calculated the successive approximations to the flow around an airfoil by means of a electrolyte tank. They found that when the maximum velocity in the flow reaches the local velocity of sound, the convergence of the successive steps broke down. This fact led the definition of critical speed or Mach number as the Mach number of the undisturbed flow for which the local velocity at some point of the surface reaches the local velocity of sound. However there is no mathematical proof for the coincidence of the critical Mach number so defined and the break down of isentropic irrotational flow. Furthermore, such a definition for critical Mach number would imply that a transition from a velocity less than that of sound, or subsonic velocity, to a velocity greater than that of sound, or supersonic velocity, does not occur in isentropic irrotational flow. On the other hand, Taylor (Ref. 3) and many others ( ) found solutions for which such a transition occurs. Finally A.M. Binnie and

S. G. Hooker (Ref. 4) shown that at least for the case of spiral flow the method of successive approximation is a convergent one even for supersonic velocities. With these fact in mind, the identification of critical speed with local supersonic velocity cannot be correct.

The first hope of clearing up this problem appeared in an investigation made by W. Tollmien (Ref. 5). In this paper, the existance of the "limiting line" or the line beyond which isentropic irrotational flow cannot continue, was first shown. Such lines was found for two-dimensional flow in a curved channel or spiral flow. The velocity at the limiting line is always supersonic. However, the true characteristics of such limiting lines + their significance were not investigated by Tollmien at that time. Recently F. Ringleb (Ref. 6) obtained another particular solution of isentropic irrotational flow in which the maximum velocity reached is approximately twice the local sound velocity. For this flow also, a limiting line appeared beyond which the flow cannot continue. Furthermore he found the singular character of the limiting line, such as the infinite acceleration + infinite pressure gradient. Th. von Kármán (Ref. 7) shown this fact for the general two-dimensional flow. He also suggested that the limiting line is the envelope of the Mach waves (Fig. 1) + thus can only occur in supersonic region. This general two-dimensional theory was established later by both Ringleb (Ref. 8) and Tollmien (Ref. 9). Tollmien corrected some mistakes appeared in Ringleb's paper and in addition, shown that the flow definitely cannot continue beyond the limiting line. The later fact introduces a "forbidden region" in the flow bounded by the limiting line. This physical absurdity can only be avoided by relaxing the condition of

He also take its appearance as the criterion for break down of isentropic irrotational flow.

4

But as stated previously, for non-viscous fluids, the transition from a flow without vorticity to that with vorticity can only be accomplished by shock waves, which at the same time also cause an increase in the entropy.

Before one can conclude that the appearance of limiting line, or the envelope of Mach waves, is the general condition for break down of isentropic irrotational flow, one must prove that the singular behavior of limiting lines are general, & not limited to two-dimensional flow. This is the purpose of the present paper. First the property of limiting line in axially symmetric flow will be investigated in detail. Then the general three dimensional problem will be sketched. It will be seen that the results of Ringleb, von Kármán and Tollmien are reconfirmed for these more general cases.

Therefore by considering only the steady flow of non-viscous fluids, the criterion for break down of isentropic irrotational flow is the appearance of limiting line. However, for the actual motion of a solid body, the flow is neither steady nor non-viscous. Small disturbances always occur and almost all real fluids have appreciable viscosity. The small disturbances in the flow introduce the question of stability. In other words, the solution found for isentropic irrotational flow may be unstable even before the appearance of limiting line, and tends to transform itself to a rotational flow involving shock waves at the slightest disturbance. If this is the case, the criterion concerns not the limiting line but the stability limit. This problem has yet to be solved.

The effect of viscosity will be limited to the boundary layer if the pressure along the surface in the flow direction never increases too rapidly. Then outside the boundary layer the flow is isentropic & irrotational. If the gradient of pressure is too large, the boundary layer will separate from the surface.

However at low velocities such separation only widens the wake of the body and change the pressure distribution over the body. But if the boundary layer separates at a point where the velocity outside the boundary layer is supersonic, additional effects may appear. The flow outside the boundary layer in this case can be regarded approximately as that of a solid body not of original contour but of a new contour including the "dead water" region created by the separation. It is then immediately clear that the ideal isentropic irrotational flow around this new contour may have a limiting line. Hence the actual flow then must involve shock waves. In other words, the separation of boundary layer in supersonic region may induce a shock wave and thus extends its influence far beyond the region of separation. Furthermore, the steep adverse pressure gradient across a shock wave may accent the separation. This interaction between the separation & the shock wave is frequently observed in experiments.

The above considerations indicate the possibility of the break down of isentropic irrotational flow outside the boundary layer even before the appearance of limiting line. Therefore the Mach number of the undisturbed flow at which the limiting line appear may be called as the "upper critical" Mach number. On the other hand, since shock wave can only occur in supersonic flow, the Mach number of the undisturbed flow at which local velocity reaches the velocity of sound may be called as the "lower critical" Mach number. The actual critical Mach number for the appearance of shock waves & the compressibility burble must lie between these two limits. By carefully design of the contour of the body to avoid the crowding together of Mach waves to form an envelope and to eliminate adverse pressure gradient along the surface of the body, the compressibility bubble can be delayed.

Axially Symmetric Flow

The solution of the exact differential equations for an axially symmetric isentropic irrotational flow was first given by F. Frankl (Ref. 10). The method is developed independently by C. Ferrari (Ref. 11). Their method applies particularly to the case of supersonic flow over a body of revolution with pointed nose. In this case, the flow at the nose can be approximated by the well-known solution for a cone. From this solution, the differential equation is solved step by step using the net of characteristics which are real for supersonic velocities. In the following investigation, the chief concern is not the solution of the partial differential equation but rather the occurance and the properties of the limiting line in *an isentropic irrotational flow. The general plan of attack is that of Tollmien (Ref. 11). However here the calculation is based on the Legendre transformation of velocity potential instead of the stream function.*

If q is the magnitude of the velocity, a the corresponding velocity of sound assuming isentropic process, p the pressure, and ρ the density of fluid, the Bernoulli equation gives

$$\frac{\rho}{\rho_0} = \left(1 - \frac{\gamma-1}{2}\frac{q^2}{a_0^2}\right)^{\frac{1}{\gamma-1}} = \left(1 + \frac{\gamma-1}{2}\frac{q^2}{a^2}\right)^{-\frac{1}{\gamma-1}} \qquad (1)$$

$$\frac{a^2}{a_0^2} = 1 - \frac{\gamma-1}{2}\frac{q^2}{a_0^2} = \left(1 + \frac{\gamma-1}{2}\frac{q^2}{a^2}\right)^{-1} \qquad (2)$$

$$\frac{p}{p_0} = \left(1 - \frac{\gamma-1}{2}\frac{q^2}{a_0^2}\right)^{\frac{\gamma}{\gamma-1}} = \left(1 + \frac{\gamma-1}{2}\frac{q^2}{a^2}\right)^{-\frac{\gamma}{\gamma-1}} \qquad (3)$$

In these equations, the subscript 0 denotes quantities corresponding to $q=0$, and γ is the ratio of specific heats of the fluids. Let the axis of symmetry be the x-axis, the distance normal to x-axis be denoted by y, and the velocity

The x-y plane is, therefore, a meridian plane

components along these two directions be denoted by u and v respectively (Fig. 2). Then the kinematical relations of the flow are given by the vorticity equation

$$\frac{\partial v}{\partial x} - \frac{\partial u}{\partial y} = 0 \qquad (4)$$

and the continuity equation

$$\frac{\partial}{\partial x}\left(y \frac{\rho}{\rho_0} u\right) + \frac{\partial}{\partial y}\left(y \frac{\rho}{\rho_0} v\right) = 0 \qquad (5)$$

Eqs (1) (2) (3) (4) and (5) specify the flow completely with the relation $q^2 = u^2 + v^2$.

To simplify the problem, a velocity potential φ defined as follows is introduced:

$$u = \frac{\partial \varphi}{\partial x}, \qquad v = \frac{\partial \varphi}{\partial y} \qquad (6)$$

Then Eq. (4) is identically satisfied and Eq. (5) together with Eqs. (1) and (2) gives the equation for φ.

$$\left(1 - \frac{u^2}{a^2}\right)\frac{\partial^2 \varphi}{\partial x^2} - 2\frac{uv}{a^2}\frac{\partial^2 \varphi}{\partial x \partial y} + \left(1 - \frac{v^2}{a^2}\right)\frac{\partial^2 \varphi}{\partial y^2} + \frac{v}{y} = 0 \qquad (7)$$

The characteristics of this differential equation, to be called the characteristics in physical plane, is given by $g(x, y) = 0$, where $g(x, y)$ is determined by the following equation

$$\left(1 - \frac{u^2}{a^2}\right)\left(\frac{\partial g}{\partial x}\right)^2 - 2\frac{uv}{a^2}\frac{\partial g}{\partial x}\frac{\partial g}{\partial y} + \left(1 - \frac{v^2}{a^2}\right)\left(\frac{\partial g}{\partial y}\right)^2 = 0 \qquad (8)$$

It can be easily seen from this equation that g is real only when $q^2 > a^2$. Therefore the characteristics are real only in supersonic regions of the flow.

The meaning of characteristics in physical plane is immediately clear if one calculates the relation between the slope of a characteristic and the slope of a stream line in the meridian or xy plane. By the definition of the function $g(x, y)$, the value of g is zero, or constant, along a characteristic. Therefore by writing a quantity evaluated at a certain constant value of a parameter with that parameter as a subscript, the slope of characteristic in physical plane is

$$\left(\frac{dy}{dx}\right)_g = -\frac{\frac{\partial g}{\partial x}}{\frac{\partial g}{\partial y}} \qquad (9)$$

Along a stream line, the stream function ψ defined by following equation is constant:

$$\frac{\partial \psi}{\partial y} = y\frac{\rho}{\rho_0}u, \quad -\frac{\partial \psi}{\partial x} = y\frac{\rho}{\rho_0}v \qquad (10)$$

Therefore the slope of a stream line is

$$\left(\frac{dy}{dx}\right)_\psi = \frac{v}{u} \qquad (11)$$

Eqs (8), (9) & (11) give

$$\left(\frac{dy}{dx}\right)_g = \frac{-\frac{uv}{a^2} \pm \sqrt{\frac{q^2}{a^2}-1}}{1-\frac{u^2}{a^2}} = \left\{\left(\frac{dy}{dx}\right)_\psi \pm \tan\beta\right\} \div \left\{1 \mp \left(\frac{dy}{dx}\right)_\psi \tan\beta\right\} \qquad (12)$$

where β is the Mach angle given by $\beta = \sin^{-1}\frac{a}{q}$. Therefore Eq. (12) shows that the characteristics in physical plane are inclined to the stream lines by an angle equal to the Mach angle. Such lines are the wave fronts of infinitesimal disturbances and are called Mach waves. In other words, characteristics in physical plane are the Mach waves in that plane. Generally there are two families of Mach waves inclined symmetrically with respect to each stream line. (Fig. 1).

If to each pair of values of u and v, there is <u>one</u> pair of values of x, y, then x and y can be considered as functions of u, v. In other words, instead of taking x and y as independent variables, u, v can be used as independent variable. The plane with u, & v as coördinates is called the "hodograph plane." An equation in hodograph plane corresponding to Eq. (7) for φ can be obtained by means of Legendre's transformation. By writing

$$\chi = ux + vy - \varphi \tag{13}$$

it is seen that

$$\frac{\partial \chi}{\partial u} = x \,, \qquad \frac{\partial \chi}{\partial v} = y \tag{14}$$

Then Eq. (7) can be written as

$$\left(1-\frac{u^2}{a^2}\right)\frac{\partial^2\chi}{\partial v^2} + 2\,\frac{uv}{a^2}\,\frac{\partial^2\chi}{\partial u\partial v} + \left(1-\frac{v^2}{a^2}\right)\frac{\partial^2\chi}{\partial u^2} + \frac{v}{\frac{\partial\chi}{\partial v}}\left[\frac{\partial^2\chi}{\partial u^2}\frac{\partial^2\chi}{\partial v^2} - \left(\frac{\partial^2\chi}{\partial u\partial v}\right)^2\right] = 0 \tag{15}$$

The characteristics of Eq. (15) are given by $f(u,v)=0$ where f is the solution of following differential equation

$$\left\{\left(1-\frac{u^2}{a^2}\right) + \frac{v}{\frac{\partial\chi}{\partial v}}\,\frac{\partial^2\chi}{\partial u^2}\right\}\left(\frac{\partial f}{\partial v}\right)^2 + 2\left(\frac{uv}{a^2} - \frac{v}{\frac{\partial\chi}{\partial v}}\,\frac{\partial^2\chi}{\partial u\partial v}\right)\frac{\partial f}{\partial v}\,\frac{\partial f}{\partial u}$$
$$+ \left\{\left(1-\frac{v^2}{a^2}\right) + \frac{v}{\frac{\partial\chi}{\partial v}}\,\frac{\partial^2\chi}{\partial v^2}\right\}\left(\frac{\partial f}{\partial u}\right)^2 = 0 \tag{16}$$

Eq. (16) shows that the characteristics in hodograph plane depends upon the values of the derivatives χ which must be obtained from Eq. (15). In other words, the characteristics in hodograph plane change with the flow and are not a constant set of curves as those in two-dimensional problems.

To obtain the relation between the characteristics in physical plane and those in hodograph plane, it is noticed that Eq. (9) can be rewritten as

$$(dy)_\varphi : (dx)_\varphi = -\frac{\partial\varphi}{\partial x} : \frac{\partial\varphi}{\partial y} \tag{17}$$

Then Eq (8) is equivalent to

$$\left(1-\frac{u^2}{a^2}\right)(dy)_\varphi^2 + 2\,\frac{uv}{a^2}\,(dy)_\varphi(dx)_\varphi + \left(1-\frac{v^2}{a^2}\right)(dx)_\varphi^2 = 0 \tag{18}$$

However, in general, Eq. (14) gives the following relation between the differentials of x & y and those of u & v:

$$dx = \frac{\partial^2 \chi}{\partial u^2} du + \frac{\partial^2 \chi}{\partial u \partial v} dv$$
$$dy = \frac{\partial^2 \chi}{\partial u \partial v} du + \frac{\partial^2 \chi}{\partial v^2} dv \qquad (19)$$

By means of these relations, Eq. (18) can be transformed into an equation for $(du)_g$ and $(dv)_g$. This transformed equation can be simplified by using Eq. (15), the final relation is

$$\left[\frac{\partial^2 \chi}{\partial u^2}\frac{\partial^2 \chi}{\partial v^2} - \left(\frac{\partial^2 \chi}{\partial u \partial v}\right)^2\right]\left[\left\{\left(1-\frac{u^2}{a^2}\right) + \frac{v}{\frac{\partial \chi}{\partial v}}\frac{\partial^2 \chi}{\partial u^2}\right\}(du)_g^2 - 2\left\{\frac{uv}{a^2} - \frac{v}{\frac{\partial \chi}{\partial v}}\frac{\partial^2 \chi}{\partial u \partial v}\right\}\right.$$
$$\left.(du)_g \cdot (dv)_g + \left\{\left(1-\frac{v^2}{a^2}\right) + \frac{v}{\frac{\partial \chi}{\partial v}}\frac{\partial^2 \chi}{\partial v^2}\right\}(dv)_g^2\right] = 0 \qquad (20)$$

Therefore if the first factor of Eq. (20) is not zero, the variations $(du)_g$ + $(dv)_g$ corresponding to a characteristic in physical plane must satisfy the relation

$$\left\{\left(1-\frac{u^2}{a^2}\right) + \frac{v}{\frac{\partial \chi}{\partial v}}\frac{\partial^2 \chi}{\partial u^2}\right\}(du)_g^2 - 2\left\{\frac{uv}{a^2} - \frac{v}{\frac{\partial \chi}{\partial v}}\frac{\partial^2 \chi}{\partial u \partial v}\right\}(du)_g (dv)_g$$
$$+ \left\{\left(1-\frac{v^2}{a^2}\right) + \frac{v}{\frac{\partial \chi}{\partial v}}\frac{\partial^2 \chi}{\partial v^2}\right\}(dv)_g^2 = 0 \qquad (21)$$

This is the same relation for the variations $(du)_f$ + $(dv)_f$ along a characteristics in hodograph plane as can be seen from Eq. (16) and the following relation obtained from the definition of f

$$(dv)_f : (du)_f = -\frac{\partial f}{\partial u} : \frac{\partial f}{\partial v} \qquad (22)$$

The transformed characteristics of physical plane (in hodograph plane) and the characteristics of hodograph plane itself, satisfy then the same first order differential equation. Therefore these two type of curves are the same. In other words, the characteristics of hodograph plane are the representation of Mach waves in u, v plane.

The Limiting Line

Eq. (20) shows that if

$$\frac{\partial^2\chi}{\partial u^2}\frac{\partial^2\chi}{\partial v^2}-\left(\frac{\partial^2\chi}{\partial u\partial v}\right)^2=0 \qquad (23)$$

then the transformed differential equation for characteristics of physical plane, or Mach waves is satisfied. Therefore if there is a line in the hodograph plane, along which the values of the derivatives of χ are such that Eq. (23) is true, then this line when transferred to the physical plane will have its slope equal to that of one family of Mach waves. Such lines are called as limiting hodograph in u-v plane and limiting line in physical plane. The significance of the adjective "limiting" will be made clear as other properties of such lines are investigated.

> It is then evident that the limiting line can only occur in supersonic region.

> Now the question arises: Can the limiting hodograph be the characteristics in u-v plane?

Along a limiting hodograph, Eq. (23) gives by differentiation

$$\left(\frac{dv}{du}\right)_l=-\frac{\dfrac{\partial^3\chi}{\partial u^3}\dfrac{\partial^2\chi}{\partial v^2}-2\dfrac{\partial^2\chi}{\partial u\partial v}\dfrac{\partial^3\chi}{\partial u^2\partial v}+\dfrac{\partial^2\chi}{\partial u^2}\dfrac{\partial^3\chi}{\partial u\partial v^2}}{\dfrac{\partial^3\chi}{\partial u^2\partial v}\dfrac{\partial^2\chi}{\partial v^2}-2\dfrac{\partial^2\chi}{\partial u\partial v}\dfrac{\partial^3\chi}{\partial u\partial v^2}+\dfrac{\partial^2\chi}{\partial u^2}\dfrac{\partial^3\chi}{\partial v^3}} \qquad (24)$$

where the subscript l denotes the value along a limiting hodograph. Now the general differential equation for χ, Eq. (15), is true for the whole u-v plane, therefore the equation is still true by differentiating it with respect to u and v. The results can be simplified by using Eq. (15) itself & Eq. (23). Then at the limiting hodograph,

$$\left[\left(1-\frac{v^2}{a^2}\right)+\frac{v}{\frac{\partial\chi}{\partial v}}\frac{\partial^2\chi}{\partial v^2}\right]\frac{\partial^3\chi}{\partial u^3}+2\left[\frac{uv}{a^2}-\frac{v}{\frac{\partial\chi}{\partial v}}\frac{\partial^2\chi}{\partial u\partial v}\right]\frac{\partial^3\chi}{\partial u^2\partial v}+\left[\left(1-\frac{u^2}{a^2}\right)+\frac{v}{\frac{\partial\chi}{\partial v}}\frac{\partial^2\chi}{\partial u^2}\right]\frac{\partial^3\chi}{\partial u\partial v^2}$$
$$=(\gamma+1)\frac{u}{a^2}\frac{\partial^2\chi}{\partial v^2}-2\frac{v}{a^2}\frac{\partial^2\chi}{\partial u\partial v}+(\gamma-1)\frac{u}{a^2}\frac{\partial^2\chi}{\partial u^2} \qquad (25a)$$

$$\left[\left(1-\frac{v^2}{a^2}\right)+\frac{v}{\frac{\partial\chi}{\partial v}}\frac{\partial^2\chi}{\partial v^2}\right]\frac{\partial^3\chi}{\partial u\partial v^2}+2\left[\frac{uv}{a^2}-\frac{v}{\frac{\partial\chi}{\partial v}}\frac{\partial^2\chi}{\partial u\partial v}\right]\frac{\partial^3\chi}{\partial u\partial v^2}+\left[\left(1-\frac{u^2}{a^2}\right)+\frac{v}{\frac{\partial\chi}{\partial v}}\frac{\partial^2\chi}{\partial u^2}\right]\frac{\partial^3\chi}{\partial v^3}$$
$$=(\gamma-1)\frac{v}{a^2}\frac{\partial^2\chi}{\partial v^2}-2\frac{u}{a^2}\frac{\partial^2\chi}{\partial u\partial v}+(\gamma+1)\frac{v}{a^2}\frac{\partial^2\chi}{\partial u^2} \qquad (25b)$$

Eqs (24), (25a), & (25b) are the only available equations involving no higher derivative than the third. On the other hand, the slope of a characteristic in hodograph plane can be calculated by Eq. (22),

$$\left(\frac{dv}{du}\right)_f = - \frac{\frac{\partial f}{\partial u}}{\frac{\partial f}{\partial v}} \tag{26}$$

This equation together with Eq. (16) gives

$$\left\{\left(1-\frac{v^2}{a^2}\right)+\frac{\frac{v}{\partial X}}{\partial v}\frac{\partial^2 X}{\partial v^2}\right\}\left(\frac{dv}{du}\right)_f^2 - 2\left\{\frac{uv}{a^2}-\frac{\frac{v}{\partial X}}{\partial v}\frac{\partial^2 X}{\partial u \partial v}\right\}\left(\frac{dv}{du}\right)_f + \left\{\left(1-\frac{u^2}{a^2}\right)+\frac{\frac{v}{\partial X}}{\partial v}\frac{\partial^2 X}{\partial u^2}\right\}=0 \tag{27}$$

Therefore if the limiting hodograph is a characteristic, then $\left(\frac{dv}{du}\right)_l$ must satisfy Eq. (27). However a simple calculation shows that it is not even possible to obtain a relation between $\left(\frac{dv}{du}\right)_l$ & other quantities not involving the third order derivatives of X. Hence $\left(\frac{dv}{du}\right)_l$ does not satisfy Eq. (27). In other words, the limiting hodograph is not a characteristic. Transferred to physical plane, this means that the limiting line is not a Mach wave. But as shown in previous paragraph, the limiting line is everywhere tagent to one family of Mach waves, Consequently, the limiting line must be the envelope of a family of Mach waves. This property of limiting line can be taken as its physical definition.

Limiting Hodograph and the Stream Lines

At the limiting hodograph both Eqs (15) and (23) holds, by eliminating one of the second order derivatives, say $\frac{\partial^2 X}{\partial u^2}$, the following relation is obtained

$$\left(\frac{\partial^2 X}{\partial v^2}\right)_l = \frac{-\frac{uv}{a^2} \pm \sqrt{\frac{q^2}{a^2}-1}}{1-\frac{u^2}{a^2}}\left(\frac{\partial^2 X}{\partial u \partial v}\right)_l \tag{28}$$

The sign before the radical in Eq. (28) can be either positive or negative but ⟷ not both. This relation will be used presently to show that the stream lines and one family of characteristics are tangent in u-v plane.

From Eq. (10), the differential of stream function can be calculated as

$$d\psi = -y\frac{\rho}{\rho_0}v\,dx + y\frac{\rho}{\rho_0}u\,dy \qquad (29)$$

In this equation, y can be replaced by $\frac{\partial\chi}{\partial v}$ according to Eq. (14) + the differentials dx + dy replaced by the differential du + dv according to Eq. (19). Then

$$d\psi = \frac{\partial\chi}{\partial v}\frac{\rho}{\rho_0}\left[\left(-v\frac{\partial^2\chi}{\partial u^2} + u\frac{\partial^2\chi}{\partial u\partial v}\right)du + \left(-v\frac{\partial^2\chi}{\partial u\partial v} + u\frac{\partial^2\chi}{\partial v^2}\right)dv\right] \qquad (30)$$

Along a stream line, $d\psi=0$, therefore the slope of stream line in hodograph plane is given by

$$\left(\frac{dv}{du}\right)_\psi = \frac{v\frac{\partial^2\chi}{\partial u^2} - u\frac{\partial^2\chi}{\partial u\partial v}}{-v\frac{\partial^2\chi}{\partial u\partial v} + u\frac{\partial^2\chi}{\partial v^2}} \qquad (31)$$

At the limiting hodograph, Eq. (23) holds, therefore Eq. (31) together with Eq. (28) gives

$$\left(\frac{dv}{du}\right)_{\psi,l} = -\left(\frac{\frac{\partial^2\chi}{\partial u\partial v}}{\frac{\partial^2\chi}{\partial v^2}}\right)_l = \frac{1-\frac{u^2}{a^2}}{\frac{uv}{a^2} \mp\sqrt{\frac{q^2}{a^2}-1}} \qquad (32)$$

where the sign before the radical can be either negative or positive corresponding to the sign in Eq. (28).

On the other hand, the slope of the characteristics in hodograph plane is determined by Eq. (27). By solving for $\left(\frac{dv}{du}\right)_f$ + simplifying the result with aid of Eq. (15),

$$\left(\frac{dv}{du}\right)_f = \frac{\frac{uv}{a^2} - \frac{v}{\frac{\partial\chi}{\partial v}}\frac{\partial^2\chi}{\partial u\partial v} \pm\sqrt{\frac{q^2}{a^2}-1}}{\left(1-\frac{v^2}{a^2}\right) + \frac{v}{\frac{\partial\chi}{\partial v}}\frac{\partial^2\chi}{\partial v^2}} \qquad (33)$$

The sign before the radical is either positive or negative corresponds to the two families of characteristics. By using the positive sign in conjunction with positive sign in Eq. (28), and similarly for the negative sign,

$$\left(\frac{dv}{du}\right)_{f,\,l} = \frac{1-\frac{u^2}{a^2}}{\frac{uv}{a^2} \mp \sqrt{\frac{q^2}{a^2}-1}} \qquad (34)$$

Eqs. (32) and (34) show that the stream-lines and one family of characteristics are tangent to each other at the limiting hodograph. This result is same as that obtained for two-dimensional flow. (Ref. 7, 8, 9). These equations when compared with Eq. (12) for the slope of Mach waves in physical plane yields the interesting result that the stream lines + one family of characteristics (at limiting hodograph) are perpendicular to a family of (corresponding) Mach waves at the limiting line

since

$$\left(\frac{dv}{du}\right)_{\psi} = -\frac{\frac{\partial\psi}{\partial u}}{\frac{\partial\psi}{\partial v}} \qquad (35)$$

Eq. (32) gives the following equation which holds at the limiting hodograph

$$\left(1-\frac{v^2}{a^2}\right)\left(\frac{\partial\psi}{\partial u}\right)_l^2 + 2\,\frac{uv}{a^2}\left(\frac{\partial\psi}{\partial u}\right)_l\left(\frac{\partial\psi}{\partial v}\right)_l + \left(1-\frac{u^2}{a^2}\right)\left(\frac{\partial\psi}{\partial v}\right)_l^2 = 0 \qquad (36)$$

This equation can be reduced to more familiar form by introducing the polar coördinates in u, v plane;

$$u = q\cos\theta, \qquad v = q\sin\theta$$

where θ is the angle between the velocity vector + x-axis. Then Eq. (36) takes the form

$$\left(\frac{\partial\psi}{\partial q}\right)_l^2 + \left(\frac{1}{q^2}-\frac{1}{a^2}\right)\left(\frac{\partial\psi}{\partial\theta}\right)_l^2 = 0 \qquad (37)$$

This can be regarded as the equivalent to Eq. (23) for defining the limit hodograph. Similar relation exists for two dimensional flow. (Ref. 7, 8, 9).

Along a stream line, the ratio between $(dv)_\psi$ & $(du)_\psi$ is given by Eq. (30). By substituting this ratio into Eq. (19), the differential $(dx)_\psi$ & $(dy)_\psi$ along a stream line is given as

$$(dx)_\psi = \frac{u\left[\frac{\partial^2\chi}{\partial u^2}\frac{\partial^2\chi}{\partial v^2} - \left(\frac{\partial^2\chi}{\partial u\partial v}\right)^2\right]}{-v\frac{\partial^2\chi}{\partial u\partial v} + u\frac{\partial^2\chi}{\partial v^2}}(du)_\psi$$

$$(dy)_\psi = \frac{v\left[\frac{\partial^2\chi}{\partial u^2}\frac{\partial^2\chi}{\partial v^2} - \left(\frac{\partial^2\chi}{\partial u\partial v}\right)^2\right]}{-v\frac{\partial^2\chi}{\partial u\partial v} + u\frac{\partial^2\chi}{\partial v^2}}(du)_\psi \qquad (38)$$

At the limiting line, Eq. (23) is satisfied. Then Eq. (38) shows that at the limiting line, the stream line has a singularity. Or, more plainly, $(dx)_\psi$ & $(dy)_\psi$ at these points are zero. By writing s for the distance measured along a stream line, Eq. (38) gives immediately

$$\left(\frac{\partial u}{\partial s}\right)_\psi = \frac{-v\frac{\partial^2\chi}{\partial u\partial v} + u\frac{\partial^2\chi}{\partial v^2}}{q\left[\frac{\partial^2\chi}{\partial u^2}\frac{\partial^2\chi}{\partial v^2} - \left(\frac{\partial^2\chi}{\partial u\partial v}\right)^2\right]} \qquad (39)$$

Similarly

$$\left(\frac{\partial v}{\partial s}\right)_\psi = \frac{v\frac{\partial^2\chi}{\partial u^2} - u\frac{\partial^2\chi}{\partial u\partial v}}{q\left[\frac{\partial^2\chi}{\partial u^2}\frac{\partial^2\chi}{\partial v^2} - \left(\frac{\partial^2\chi}{\partial u\partial v}\right)^2\right]} \qquad (40)$$

Therefore at the limiting line, the accelerations along a stream line is infinitely large. Furthermore since the pressure gradient $\left(\frac{\partial p}{\partial s}\right)_\psi$ along a stream line is

$$\left(\frac{\partial p}{\partial s}\right)_\psi = -\rho q\frac{\partial q}{\partial s} = -\rho\left[u\left(\frac{\partial u}{\partial s}\right)_\psi + v\left(\frac{\partial v}{\partial s}\right)_\psi\right] \qquad (41)$$

the pressure gradient at the limiting line is also infinitely large.

Such infinite acceleration and pressure gradient lead one to suspect that the fluid is thrown back at the limiting line. In other words, the stream lines are doubled back at this line of singularity. To investigate whether this is true, the character of the relation $\frac{\partial^2 \chi}{\partial u^2}\frac{\partial^2 \chi}{\partial v^2} - \left(\frac{\partial^2 \chi}{\partial u \partial v}\right)^2 = 0$ along a stream line has to be determined. If the derivative of this expression along a stream line is not zero, then $\frac{\partial^2 \chi}{\partial u^2}\frac{\partial^2 \chi}{\partial v^2} - \left(\frac{\partial^2 \chi}{\partial u \partial v}\right)^2$ has only a simple zero at the intersection of the limiting line and the stream line. Consequently, the differentials $(dx)_\psi$ & $(dy)_\psi$ will change sign by passing through the limiting hodograph in u-v plane along a stream line. Hence the stream lines will double back and form a cusp at the limiting line. The derivative of $\frac{\partial^2 \chi}{\partial u^2}\frac{\partial^2 \chi}{\partial v^2} - \left(\frac{\partial^2 \chi}{\partial u \partial v}\right)^2$ along the stream line can be calculated with the aid of Eq. (30)

$$\frac{d}{du}\left[\frac{\partial^2 \chi}{\partial u^2}\frac{\partial^2 \chi}{\partial v^2} - \left(\frac{\partial^2 \chi}{\partial u \partial v}\right)^2\right]_\psi = \frac{\partial^3 \chi}{\partial u^3}\frac{\partial^2 \chi}{\partial v^2} + \frac{\partial^2 \chi}{\partial u^2}\frac{\partial^3 \chi}{\partial u \partial v^2} - 2\frac{\partial^2 \chi}{\partial u \partial v}\frac{\partial^3 \chi}{\partial u^2 \partial v}$$

$$+ \frac{v\frac{\partial^2 \chi}{\partial u^2} - u\frac{\partial^2 \chi}{\partial u \partial v}}{-v\frac{\partial^2 \chi}{\partial u \partial v} + u\frac{\partial^2 \chi}{\partial v^2}}\left\{\frac{\partial^3 \chi}{\partial u^2 \partial v}\frac{\partial^2 \chi}{\partial v^2} + \frac{\partial^2 \chi}{\partial u^2}\frac{\partial^3 \chi}{\partial v^3} - 2\frac{\partial^2 \chi}{\partial u \partial v}\frac{\partial^3 \chi}{\partial u \partial v^2}\right\} \qquad (42)$$

The expression on the right of Eq. (42) can not be reduced to zero by the available relations which consists of Eq. (23), Eq. (15) and differentiated forms of Eq. (15). Therefore the expression concerned generally only has a simple zero at the limiting hodograph and the stream lines are doubled back at the limiting line. It will be shown later that there is no solution possible beyond the limiting line. Hence the name limiting line.

Envelope of Characteristics in Hodograph Plane and Lines of Constant Velocity in Physical Plane

Since the limiting line is the envelope of the Mach waves in physical plane, it is interesting to see whether there is also envelope for the characteristics in hodograph plane. The characteristics in u-v plane are determined by Eq. (26). The envelope to them can be found by eliminating $\left(\frac{dv}{du}\right)_f$ between Eq. (26) and the following equation

$$\left\{\left(1-\frac{v^2}{a^2}\right)+\frac{v}{\frac{\partial\chi}{\partial v}}\frac{\partial^2\chi}{\partial v^2}\right\}\left(\frac{dv}{du}\right)_f-\left\{\frac{uv}{a^2}-\frac{v}{\frac{\partial\chi}{\partial v}}\frac{\partial^2\chi}{\partial u\partial v}\right\}=0 \qquad (43)$$

which is obtained by equating to zero the partial derivative of Eq. (26) with respect to $\left(\frac{dv}{du}\right)_f$. The result can be simplified by Eq. (15), and then it is simply

$$1-\frac{u^2+v^2}{a^2}+\frac{u^2v^2}{a^4}=\frac{u^2v^2}{a^4} \qquad (44)$$

This is satisfied by either

$$a=0 \qquad (45)$$

or

$$u^2+v^2=a^2. \qquad (46)$$

The first condition, Eq. (45), when substituted into Eq. (26) gives

$$\left(\frac{dv}{du}\right)_{f\,a=0}=-\frac{u}{v} \qquad (47)$$

which shows that the circle of maximum velocity corresponding to $a=0$, is the envelope to the characteristics in hodograph plane. The second condition, Eq. (46) is the spurious solution, since generally the characteristic at $q=a$ does not tangent to the circle $q=a$. Hence $a=0$ is the only envelope.

The lines of constant velocity in hodograph plane are simply circles. Therefore

$$\left(\frac{dv}{du}\right)_q=-\frac{u}{v} \qquad (48)$$

By means of this relation & Eq. (19), the slope of the lines of constant velocity is given as

$$\left(\frac{dy}{dx}\right)_q=\frac{v\frac{\partial^2\chi}{\partial u\partial v}-u\frac{\partial^2\chi}{\partial v^2}}{v\frac{\partial^2\chi}{\partial u^2}-u\frac{\partial^2\chi}{\partial u\partial v}} \qquad (49)$$

This equation together with Eq. (30) gives the following interesting relation

$$\left(\frac{dy}{dx}\right)_q = -\frac{1}{\left(\frac{dv}{du}\right)_\varphi} \tag{50}$$

In other words, a line of constant velocity in physical plane is perpendicular to the stream line in hodograph plane at corresponding point.

The Lost Solution

Throughout the previous calculation, the possibility of using the Legendre transformation is assumed. This requires that for each pair of values of u, v there is one & only one pair of values of x, y. However it is not always true, it is possible to have a number of points in the physical plane having the same value of u & v. If this is the case, then evidently it is impossible to solve for x and y from the pair of functions $u = u(x, y)$; $v = v(x, y)$. Mathematically, the situation is expressed by saying that the Jacobian $\partial(u, v)/\partial(x, y)$ vanishes in the physical plane. Or

$$\frac{\partial u}{\partial x}\frac{\partial v}{\partial y} - \frac{\partial u}{\partial y}\frac{\partial v}{\partial x} = 0 \tag{51}$$

However this is also the condition for a functional relation between u and v, e.g., v can be expressed as a function of u. In other words, u & v are not independent. Hence if a solution is "lost" or not included in the family of solutions allowing Legendre transformation, then for that solution,

$$v = v(u) \tag{52}$$

By eliminating ρ from the continuity equation, we obtain

$$\left(1-\frac{u^2}{a^2}\right)\frac{\partial u}{\partial x} - \frac{uv}{a^2}\left(\frac{\partial u}{\partial y} + \frac{\partial v}{\partial x}\right) + \left(1-\frac{v^2}{a^2}\right)\frac{\partial v}{\partial y} + \frac{v}{y} = 0 \tag{53}$$

This equation can be rewritten in the following form by using Eq. (52),

$$\left\{\left(1-\frac{u^2}{a^2}\right)-\frac{uv}{a^2}\frac{dv}{du}\right\}\frac{\partial u}{\partial x}+\left\{\left(1-\frac{v^2}{a^2}\right)\frac{dv}{du}-\frac{uv}{a^2}\right\}\frac{\partial u}{\partial y}+\frac{v}{y}=0 \qquad (54)$$

The vorticity equation, Eq. 14) can be expressed as

$$\frac{dv}{du}\frac{\partial u}{\partial x}-\frac{\partial u}{\partial y}=0 \qquad (55)$$

From Eqs (54) & (55), one can solve for $\frac{\partial u}{\partial x}$ & $\frac{\partial u}{\partial y}$. The result is

$$\left[\left(1-\frac{u^2}{a^2}\right)-2\frac{uv}{a^2}\frac{dv}{du}+\left(1-\frac{v^2}{a^2}\right)\left(\frac{dv}{du}\right)^2\right]\frac{\partial u}{\partial x}=-\frac{v}{y}$$

$$\left[\left(1-\frac{u^2}{a^2}\right)-2\frac{uv}{a^2}\frac{dv}{du}+\left(1-\frac{v^2}{a^2}\right)\left(\frac{dv}{du}\right)^2\right]\frac{\partial u}{\partial y}=-\frac{v}{y}\frac{dv}{du} \qquad (56)$$

By differentiating the first of Eq. (56) with respect to y, the second with respect to x, the following relation is obtained by subtraction:

$$\frac{d^2v}{du^2}\frac{\partial u}{\partial x}+\frac{1}{y}=0 \qquad (57)$$

Therefore
$$\frac{dv}{du}=\frac{f(y)-x}{y}$$

or
$$y=\frac{f(y)-x}{\frac{dv}{du}} \qquad (58)$$

~~where $f(y)$ is a undetermined function of y.~~

However Eq. (55) shows that for lines of constant values of u, i.e., $du=\frac{\partial u}{\partial x}(dx)_u+\frac{\partial u}{\partial y}(dy)_u=0$,

$$\left(\frac{dy}{dx}\right)_u=-\frac{1}{\left(\frac{dv}{du}\right)_u}=\text{constant} \qquad (59)$$

Hence lines of constant values of u & v are straight lines. This restriction reduces the function $f(y)$ in Eq. (58) to a numerical constant. Put $f(y)=K$, Eq. (58) is then

$$y = \frac{K - x}{\frac{dv}{du}} \qquad (60)$$

Therefore lines of constant values of u + v are radial lines passing through the point x = K. Thus the last solution is nothing but the well-known solution for the flow over a conical surface.

From Eq. (59), it is seen that lines of constant velocity is perpendicular to the tangent of u-v curve at the corresponding points. By substituting the values of $\frac{\partial u}{\partial x}$ and $\frac{\partial u}{\partial y}$ from Eq. (56) into Eq. (54), a relation between u + v is obtained:

$$v \frac{d^2 v}{du^2} - \left(1 - \frac{v^2}{a^2}\right)\left(\frac{dv}{du}\right)^2 + 2\frac{uv}{a^2}\frac{dv}{du} - \left(1 - \frac{u^2}{a^2}\right) = 0 \qquad (61)$$

This is the differential equation for the determining the hodograph representing the flow over a cone. Fig. 3 shows the hodograph for a cone of 30° semi-vertex angle and with ~~a~~ velocity at surface of cone equal to   . This is drawn from data given by G. I. Taylor and J. W. Maccoll (Ref. 12).

It may well be mentioned here that the last solution for the axially symmetric flow is not limited to supersonic velocity only as is the case for two-dimensional flow. In fact, Taylor + Maccoll show that for smaller forward velocity of the cone, the velocity just after the head shock wave is supersonic. This supersonic velocity is gradually diminished untill it becomes subsonic for points near the surface of the cone. Fig. 4 shows one example taken from their calculations (Ref. 12). Furthermore spark photographs of conical shell in actual flight taken by Maccoll (Ref. 13) do not indicate the presence of shock waves in region of flow where such transition from supersonic to subsonic velocities is expected. Therefore at least for this particular type of flow, a smooth transition through sonic velocity actually takes place.

Continuation of Solution Beyond the Limiting Line

Since it is shown in a previous paragraph that the stream lines are generally turned back at the limiting line, the question arises: Is it possible to continue the solution beyond the limiting line? Of course, there are two ways of continuing the solution; either the new solution is joined smoothly to the given solution at the limiting line or it is joined with a discontinuity. As shown before, the limiting line is the envelope of one family of the Mach waves, then at every point of this line its direction differes from that of stream line by a angle equal to the Mach angle. But the Mach angle is not zero except at points where the velocity of fluid has reached the maximum velocity and the ratio $\frac{a}{q} = 0$. Therefore the limiting line generally does not coincide with the stream line, and the discontinuity at the junction of the solution at the limiting line cannot be that of a vortex sheet. The only other type of discontinuity is the shock wave. However the angle between the limiting line & flow direction is equal to Mach angle. Then according the known result of the theory of shock waves, the discontinuity across such a line vanishes. In other words, there cannot be a discontinuity at the limiting line. Therefore it is impossible to join a new solution at the limiting line with a discontinuity.

As to the second possibility of joining a new solution smoothly at the limiting line, it is seen that the flow beyond the limiting line must be irrotational & isentropic since the limiting line cannot be a shock wave. There are only two type of isentropic irrotational flow: one that allows the Legendre transformation and one that does not, the lost solution. Investigate the second alternative first. If the solution beyond the limiting line belongs to the so-called "lost solution", then since the junction at the limiting line must be smooth, the values of u, & v at the limiting line must also satisfy the Eq. (61). But, the slope $\left(\frac{dv}{du}\right)_l$ at limiting line is given by Eq. (24). The second derivative $\left(\frac{d^2v}{du^2}\right)_l$ will then involve the fourth order derivative of X. Besides these expressions, the available relations are Eqs. (23), (15), (25a), (25b) and three more equations obtained by differentiating Eqs (25) with respect to u and v. However it is still impossible for $\left(\frac{dv}{du}\right)_l$ to satisfy an equation like Eq. (61) where no derivative of X appears. Hence the limiting hodograph does not satisfy the equation for lost solution, and the "lost solution" cannot be to continue the flow beyond the limiting line.

The only remaining possibility is to continue the flow smoothly by another solution obtainable by Legendre transformation. Smooth continuation means that the values of u, v ~~and ρ~~ must be same ~~at the~~ junction, the limiting line. ~~Since shock waves do not appear, isentropic relations still hold. The density ρ is determined by velocity only.~~ The value of u, & v are determined by the coordinate in hodograph plane. The position of the limiting line in physical is determined by $\frac{\partial X}{\partial u}$, $\frac{\partial X}{\partial v}$. Therefore the problem can be as follows: At a certain given curve $u(\lambda), v(\lambda)$ in the hodograph plane, the limiting hodograph, the values of $\frac{\partial X}{\partial u}$, $\frac{\partial X}{\partial v}$ are given (λ is the parameter along the given curve). It is required to determine a new solution of the differential equation Eq. (15) with these initial values. First of all, it is seen that with the given data, the left hand sides of the following equations are given:

$$\frac{d}{d\lambda}\left(\frac{\partial X}{\partial u}\right) = \frac{\partial^2 X}{\partial u^2}\frac{du}{d\lambda} + \frac{\partial^2 X}{\partial u \partial v}\frac{dv}{d\lambda} \qquad (62a)$$

$$\frac{d}{d\lambda}\left(\frac{\partial X}{\partial v}\right) = \frac{\partial^2 X}{\partial u \partial v}\frac{du}{d\lambda} + \frac{\partial^2 X}{\partial v^2}\frac{dv}{d\lambda} \qquad (62b)$$

therefore $$\frac{\partial^2 X}{\partial u \partial v} = -\frac{\frac{dv}{d\lambda}}{\frac{du}{d\lambda}}\frac{\partial^2 X}{\partial v^2} + \frac{1}{\frac{du}{d\lambda}}\frac{d}{d\lambda}\left(\frac{\partial X}{\partial v}\right) \qquad (63a)$$

$$\frac{\partial^2 X}{\partial u^2} = -\left(\frac{dv}{d\lambda}\right)^2\Big/\left(\frac{du}{d\lambda}\right)^2 \cdot \frac{\partial^2 X}{\partial v^2} - \frac{\frac{dv}{d\lambda}}{\left(\frac{du}{d\lambda}\right)^2}\frac{d}{d\lambda}\left(\frac{\partial X}{\partial v}\right) + \frac{1}{\frac{du}{d\lambda}}\frac{d}{d\lambda}\left(\frac{\partial X}{\partial u}\right) \qquad (63b)$$

By substituting these values into Eq. (15), the second degree terms reduces to

$$\frac{\partial^2 X}{\partial u^2}\frac{\partial^2 X}{\partial v^2} - \left(\frac{\partial^2 X}{\partial u \partial v}\right)^2 = -\left[\frac{\frac{dv}{d\lambda}}{\left(\frac{du}{d\lambda}\right)^2}\frac{d}{d\lambda}\left(\frac{\partial X}{\partial v}\right) + \frac{1}{\frac{du}{d\lambda}}\frac{d}{d\lambda}\left(\frac{\partial X}{\partial u}\right)\right]\frac{\partial^2 X}{\partial v^2}$$

$$+ \frac{1}{\left(\frac{du}{d\lambda}\right)^2}\left[\frac{d}{d\lambda}\left(\frac{\partial X}{\partial v}\right)\right]^2 \qquad (64)$$

which is linear in $\frac{\partial^2 X}{\partial v^2}$. (therefore $\frac{\partial^2 X}{\partial v^2}$ can be uniquely determined by Eq. (15)) In other words, with the given data, the second order derivatives of X at the given curve $u(\lambda)$, $v(\lambda)$ can be determined uniquely, in spite of the fact that the differential equation (15) is of second degree. K. Friedrichs and H. Lewy (Ref. 13) shown that under these circumstances, the function X within a region R (Fig. 5) bounded by two characteristics and the given curve is uniquely determined except an additional constant. Consequently there can be only one solution corresponding to the given data at the limiting hodograph. However this solution is ~~the very~~ one which gives the reverse flow at limiting line. Therefore it is impossible to continue the solution beyond the limiting line even by Legendre transformation.

Since all three alternatives ~~fail to~~ offer a way of continuing the solution, the limiting line is truely an impossible boundary to cross. In other words, the region beyond the limiting line is a "forbidden region". This physical absurdity can only be resolved by the break down of isentropic irrotational flow some distance ahead of the limiting line.

General Three Dimensional Flow

The methods used in previous section for investigating the axially symmetric flow can be easily extended to the general three dimensional case. In the present section, this investigation will be sketched and the results indicated.

Let the three components of velocity along the three coördinate axes x, y and z be denoted by u, v, and w respectively. Then by introducing a velocity potential φ defined by

$$u=\frac{\partial\varphi}{\partial x},\quad v=\frac{\partial\varphi}{\partial y},\quad w=\frac{\partial\varphi}{\partial z} \tag{65}$$

the differential equation for φ of an isentropic irrotational flow can be written as (Ref. 7)

$$a^2\left(\frac{\partial^2\varphi}{\partial x^2}+\frac{\partial^2\varphi}{\partial y^2}+\frac{\partial^2\varphi}{\partial z^2}\right)=u^2\frac{\partial^2\varphi}{\partial x^2}+v^2\frac{\partial^2\varphi}{\partial y^2}+w^2\frac{\partial^2\varphi}{\partial z^2}+2vw\frac{\partial^2\varphi}{\partial y\partial z}+2wu\frac{\partial^2\varphi}{\partial z\partial x}+2uv\frac{\partial^2\varphi}{\partial x\partial y} \tag{66}$$

If for every triad of u, v, w, there is only one triad of x, y, z, then the Legendre transformation can be used. Thus

$$\chi = ux+vy+wz-\varphi \tag{67}$$

and

$$x=\frac{\partial\chi}{\partial u},\quad y=\frac{\partial\chi}{\partial v},\quad z=\frac{\partial\chi}{\partial w} \tag{68}$$

The differential equation for φ, Eq. (66), is then transformed into

$$a^2\left[BC-F^2+CA-G^2+AB-H^2\right]=u^2(BC-F^2)+v^2(CA-G^2)+w^2(AB-H^2)+2vw(GH-AF)+2wu(HF-BG)+2uv(FG-CH) \tag{69}$$

where the following notations are used

$$A=\frac{\partial^2\chi}{\partial u^2},\quad B=\frac{\partial^2\chi}{\partial v^2},\quad C=\frac{\partial^2\chi}{\partial w^2},\quad F=\frac{\partial^2\chi}{\partial v\partial w},\quad G=\frac{\partial^2\chi}{\partial w\partial u},\quad H=\frac{\partial^2\chi}{\partial u\partial v} \tag{70}$$

By analogy with the axially symmetric case, the limiting hodograph surface is defined as the surface in the u, v, w space, or hodograph space, where the following relation holds:

$$\Delta = \begin{vmatrix} A & H & G \\ H & B & F \\ G & F & C \end{vmatrix}_{\ell} = 0 \qquad (71)$$

The properties of this limiting hodograph and the corresponding limiting surface can be found by considering the behavior of stream lines + characteristics at such surfaces.

From Eq. (68), the differentials of x, y + z can be written as

$$dx = A \cdot du + H \cdot dv + G\, dw \qquad (72a)$$

$$dy = H\, du + B\, dv + F\, dw \qquad (72b)$$

$$dz = G\, du + F\, dv + C\, dw \qquad (72c)$$

Along a stream line, the differentials dx, dy and dz must be proportional to u, v and w respectively. Thus the equation of a stream line in physical space is

$$\frac{(dx)_\psi}{u} = \frac{(dy)_\psi}{v} = \frac{(dz)_\psi}{w} \qquad (73)$$

where the subscript ψ indicates values taken along the stream line

The equation of a stream line in hodograph space is obtained by eliminating dx, dy + dz from Eq. (73) by Eq. (72). The result is

$$\frac{(du)_\psi}{\bar{a}u + \bar{h}v + \bar{g}w} = \frac{(dv)_\psi}{\bar{h}u + \bar{b}v + \bar{f}w} = \frac{(dw)_\psi}{\bar{g}u + \bar{f}v + \bar{c}w} \qquad (74)$$

where $\bar{a}$ is the cofactor of A in the determinante Δ of Eq. (71), $\bar{b}$ the cofactor of B, etc. Eq. (74) can be used in turn to eliminate two of the three differentials du, dv and dw on the right of Eq. (72). The result is

$$(dx)_\psi = \frac{u\,\Delta\, du}{\bar{a}u + \bar{h}v + \bar{g}w} \qquad (75\text{ a})$$

$$(dy)_\psi = \frac{v\,\Delta\, dv}{\bar{h}u + \bar{b}v + \bar{f}w} \qquad (75\text{ b})$$

$$(dz)_\psi = \frac{w\,\Delta\, dw}{\bar{g}u + \bar{f}v + \bar{c}w} \qquad (75\text{ c})$$

At the limiting surface, $\Delta = 0$ as defined by Eq. (71), therefore the stream lines have a singularity there. Similar to the axially symmetric flow, the stream lines generally are turned back and form a cusp at this surface. The acceleration & the pressure gradient are, of course, infinitely large at such places.

The characteristic surface (in physical space) $f(x, y, z) = 0$ is determined by the equation

$$a^2\left[\left(\frac{\partial f}{\partial x}\right)^2 + \left(\frac{\partial f}{\partial y}\right)^2 + \left(\frac{\partial f}{\partial z}\right)^2\right] = u^2\left(\frac{\partial f}{\partial x}\right)^2 + v^2\left(\frac{\partial f}{\partial y}\right)^2 + w^2\left(\frac{\partial f}{\partial z}\right)^2 + 2vw\left(\frac{\partial f}{\partial y}\right)\left(\frac{\partial f}{\partial z}\right) + 2wu\frac{\partial f}{\partial z}\frac{\partial f}{\partial x} + 2uv\left(\frac{\partial f}{\partial x}\right)\left(\frac{\partial f}{\partial y}\right) \qquad (76)$$

Since this equation is a second degree equation, there are two surfaces passing through each point. These surfaces are the wave fronts of infinitesimal disturbances in the flow and can be called the Mach surfaces. The characteristic surface $g(u, v, w) = 0$ in the hodograph space is determined by the equation

$$a^2\left[(B+C)\left(\frac{\partial g}{\partial u}\right)^2 + (C+A)\left(\frac{\partial g}{\partial v}\right)^2 + (A+B)\left(\frac{\partial g}{\partial w}\right)^2 - 2F\frac{\partial g}{\partial v}\frac{\partial g}{\partial w} - 2G\frac{\partial g}{\partial w}\frac{\partial g}{\partial u} - 2H\frac{\partial g}{\partial u}\frac{\partial g}{\partial v}\right]$$

$$= u^2\left[C\left(\frac{\partial g}{\partial v}\right)^2 + B\left(\frac{\partial g}{\partial w}\right)^2 - 2F\frac{\partial g}{\partial v}\frac{\partial g}{\partial w}\right] + v^2\left[C\left(\frac{\partial g}{\partial u}\right)^2 + A\left(\frac{\partial g}{\partial w}\right)^2 - 2G\frac{\partial g}{\partial w}\frac{\partial g}{\partial u}\right]$$

$$+ w^2\left[B\left(\frac{\partial g}{\partial u}\right)^2 + A\left(\frac{\partial g}{\partial v}\right)^2 - 2H\frac{\partial g}{\partial u}\frac{\partial g}{\partial v}\right] + 2vw\left[H\frac{\partial g}{\partial w}\frac{\partial g}{\partial u} + G\frac{\partial g}{\partial u}\frac{\partial g}{\partial v} - F\left(\frac{\partial g}{\partial u}\right)^2 - A\frac{\partial g}{\partial v}\frac{\partial g}{\partial w}\right]$$

$$+ 2wu\left[H\frac{\partial g}{\partial v}\frac{\partial g}{\partial w} + F\frac{\partial g}{\partial u}\frac{\partial g}{\partial v} - G\left(\frac{\partial g}{\partial v}\right)^2 - B\frac{\partial g}{\partial w}\frac{\partial g}{\partial u}\right]$$

$$+ 2uv\left[G\frac{\partial g}{\partial v}\frac{\partial g}{\partial w} + F\frac{\partial g}{\partial w}\frac{\partial g}{\partial u} - H\left(\frac{\partial g}{\partial w}\right)^2 - C\frac{\partial g}{\partial u}\frac{\partial g}{\partial v}\right] \qquad (77)$$

By transforming Eq. (76) for Mach surfaces to hodograph space, it can be shown that the transformed equation is satisfied by either the characteristics in hodograph space determined by Eq. (77) or the limiting hodograph determined by Eq. (71). Therefore here again the limiting surface is the envelope of a family of Mach surfaces.

By using Eqs (74) and (77), it is possible to show that the stream lines in hodograph space are tangent to the characteristic surfaces at the limiting hodograph. Furthermore by using Eqs. (69), (71) and (74), the inclination of the stream lines at the limiting hodograph can be calculated. In fact, if $ds^2 = (du)^2 + (dv)^2 + (dw)^2$, $q^2 = u^2 + v^2 + w^2$, the following relation is obtained

$$\left(\frac{ds}{dq}\right)_{\psi,l} = \frac{q}{a} \quad \text{or} \quad -\frac{q}{a} \tag{78}$$

This relation is really equivalent to Eq. (32). It thus seems the break down of general steady isentropic irrotational flow of non-viscous fluid is connected with the appearance of the envelope of Mach waves in physical space and the tangency of stream lines & characteristics in hodograph space.

In other words, the inclination of the stream lines & characteristics from the q = constant surface is equal to the Mach angle (Fig. 6). at the limiting hodograph.

References:

(2) Taylor, G.I., and Sharman, C.F., "A Mechanical Method for Solving Problems of Flow in Compressible Fluids". Proc. of Royal Society (A), Vol. 121, pp. 194 - 217, (1928)

(1) Theodorsen, T. "The Reaction on a Body in a Compressible Fluid" J. Aero. Sciences, Vol. 4, pp. 239 - 240, (1937)

(2) Lock, C.N.H., "Problems of High Speed Flight as Effected by Compressibility". J. Royal Aero. Society, Vol. 42, pp. 205 - 209, (1938)

(3) Stack, J., Lindsey, W.F., and Littell, R.E., "The Compressibility Bubble and the Effect of Compressibility on the Pressure and Forces Acting on an Airfoil". N.A.C.A. Technical Report No. 640. (1938)

(3) Taylor, G. I., "Recent Works on the Flow of Compressible Fluids" Journal London Mathe. Society Vol. 5, pp. 224 - 240. (1930)

(4) Binnie, A.M. and Hooker, S.G., "The Radial and Spiral Flow of a Compressible Fluid". Philosophical Magazine, (7), Vol 23, pp. 597-606 (1937)

(5) Tollmien, W., "Zum Übergang von Unterschall- in Überschallströmungen, Z. a. M. M., Vol. 17, pp. 117-136, (1937)

(6) Ringleb, F., "Exakte Lösungen der Differentialgleichungen einer adiabatischen Gasströmung, Ibid, Vol. 20, pp. 185 - 198, (1940)

(7) von Kármán, Th., "Compressibility Effects in Aerodynamics" J. Aero. Sciences, Vol. 8, pp. 337 - 356, in particular pp. 351 - 356, (1941)

(8) Ringleb. F., "Über die Differentialgleichungen einer adiabatischen Gasströmung und den Strömungsstoß. Deutsche Mathematik, Vol. 5, pp. 377 - 384 (1940)

(9) Tollmien, W., "Grenzlinien adiabatischer Potentialströmungen." Z. a. M. M., Vol. 21, pp. 140 - 152 (1941)

(10). Frankle, F. "

The convergence proof is given by Frankle, F. and Aleksejeva, R., "Zwei Randwertaufgaben aus der Theorie der hyperbolischen partiellen Differentialgleichungen zweiter Ordnung mit Anwendungen auf Gasströmungen mit Überschallgeschwindigkeit." Matematičeski Sbornik, Vol. 41, pp. 483 - 502, (1935) (Russian, German summary)

(13) Ferrari, C., "Campo aerodinamico a velocità iperacustica attorno a un solido di rivoluzione a prora acuminata." L'Aerotecnica, Vol. 16, pp. 121 - 130 (1936)

Also "Determinazione della pressione sopra solidi di rivoluzione a prora acuminata disposti in deriva in corrente di fluido compressibile a velocità ipersonora." Atti della R. Accademia delle Scienze di Torino, Vol. 72, pp. (1937)

(12) Taylor, G. I. and Maccoll, J. W. "The air pressure on a cone moving at high speeds." Proc. Royal Society (A), Vol. 139, pp. 278-298 (1933)

(13) Maccoll, J. W. "The Conical Shock Wave formed by a Cone moving at a High speed". Ibid. Vol. 159, pp. 459-472 (1936)

(14) Friedrichs, K. and Lewy, H. "Das Anfangswertproblem einer beliebigen nichtlinearen hyperbolischen Differentialgleichung beliebiger Ordnung in zwei Variablen." Math. Annalen, Vol. 99, pp. 200-221, (1928)

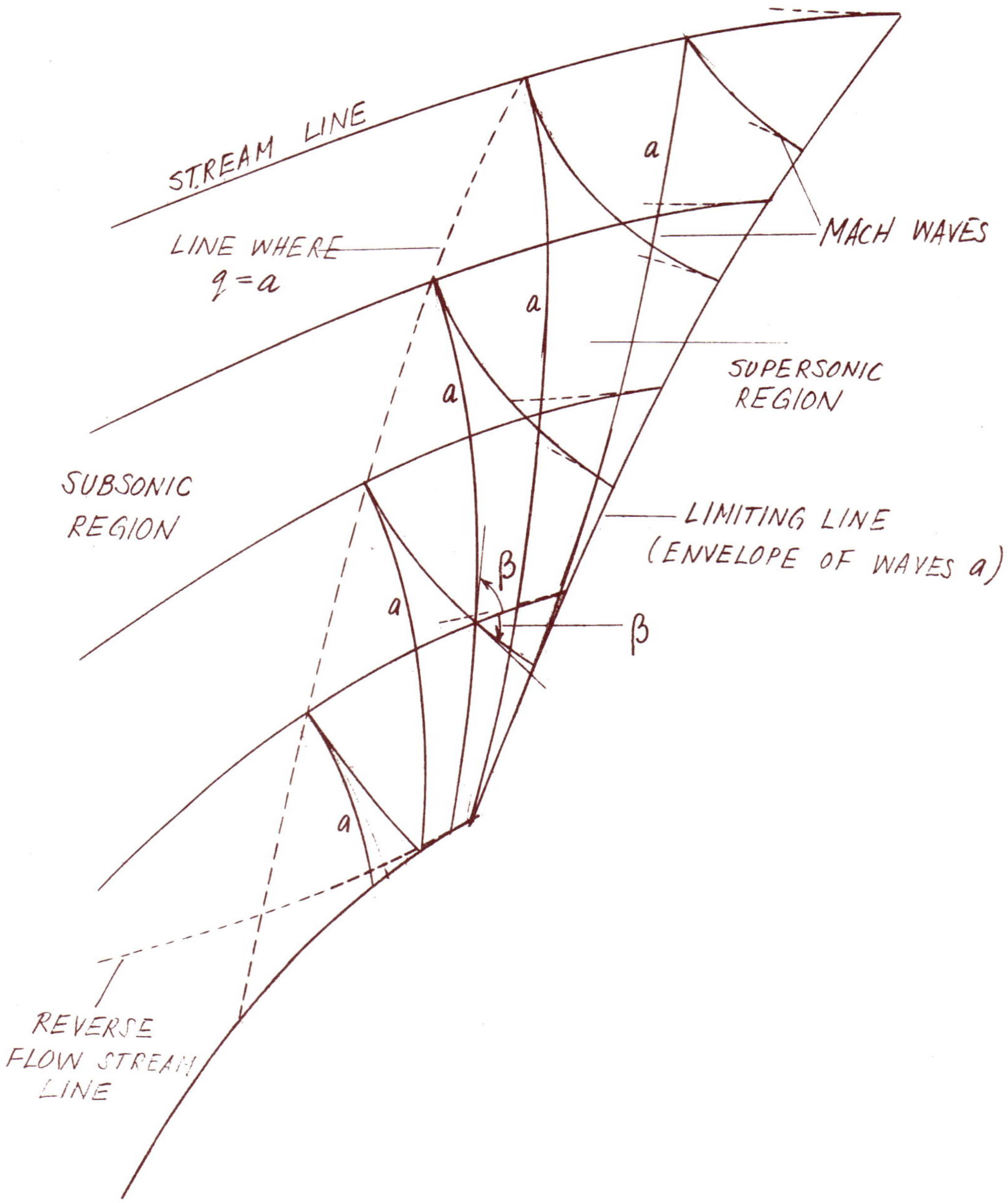

STREAM LINE
LINE WHERE
q = a
a
MACH WAVES
SUPERSONIC
REGION
SUBSONIC
REGION
LIMITING LINE
(ENVELOPE OF WAVES a)
β
REVERSE
FLOW STREAM
LINE

FIG 1

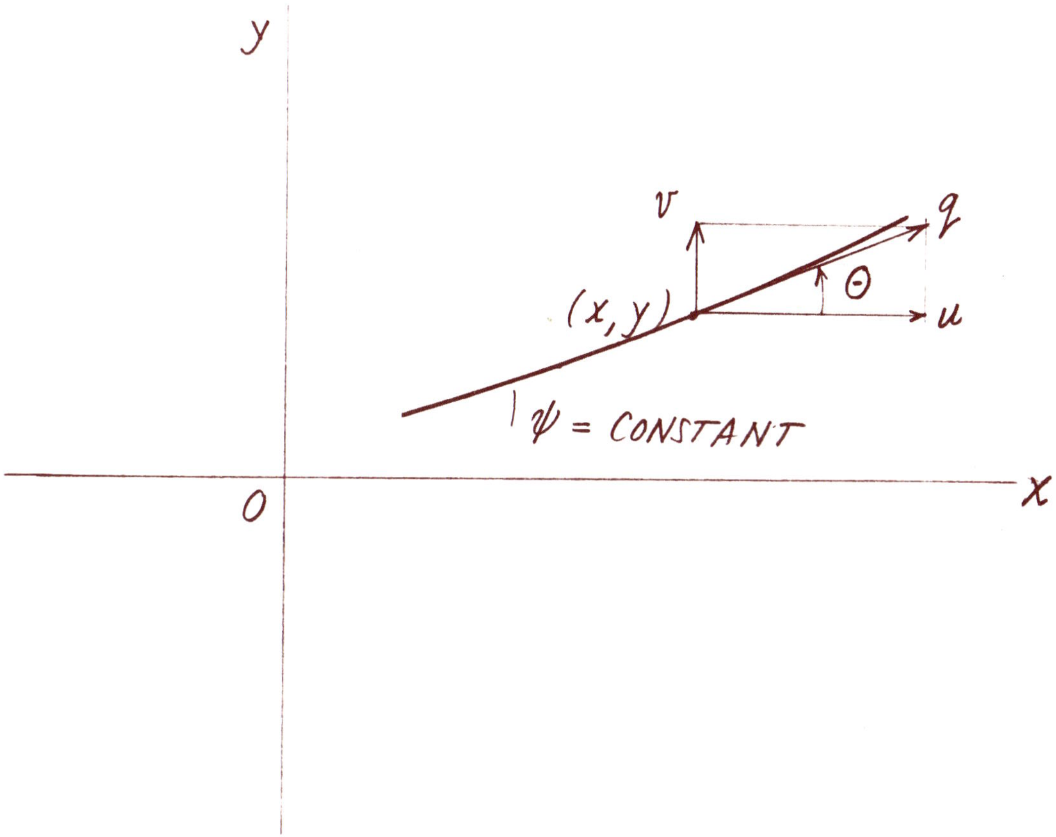

FIG.2

FIG. 3

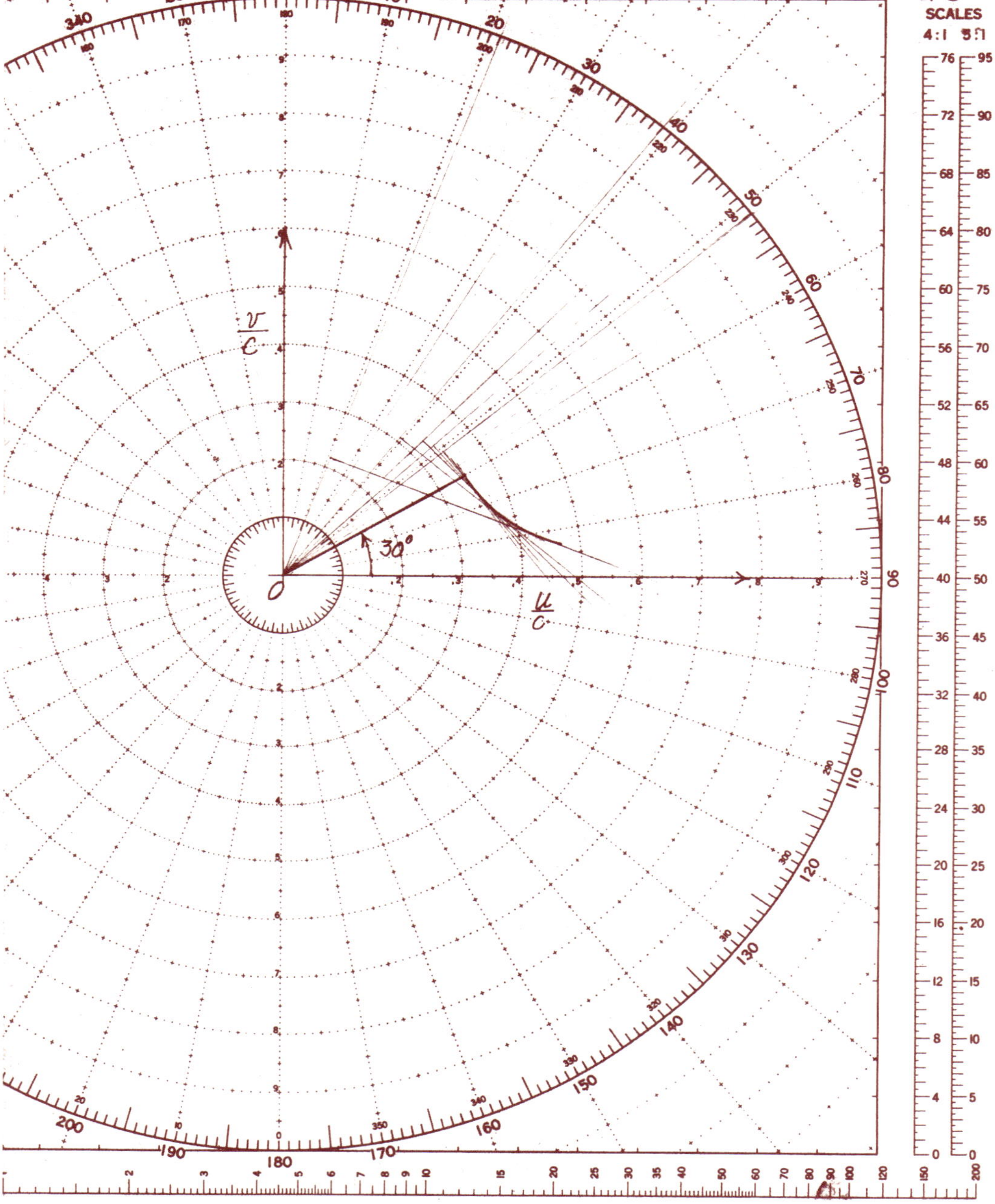

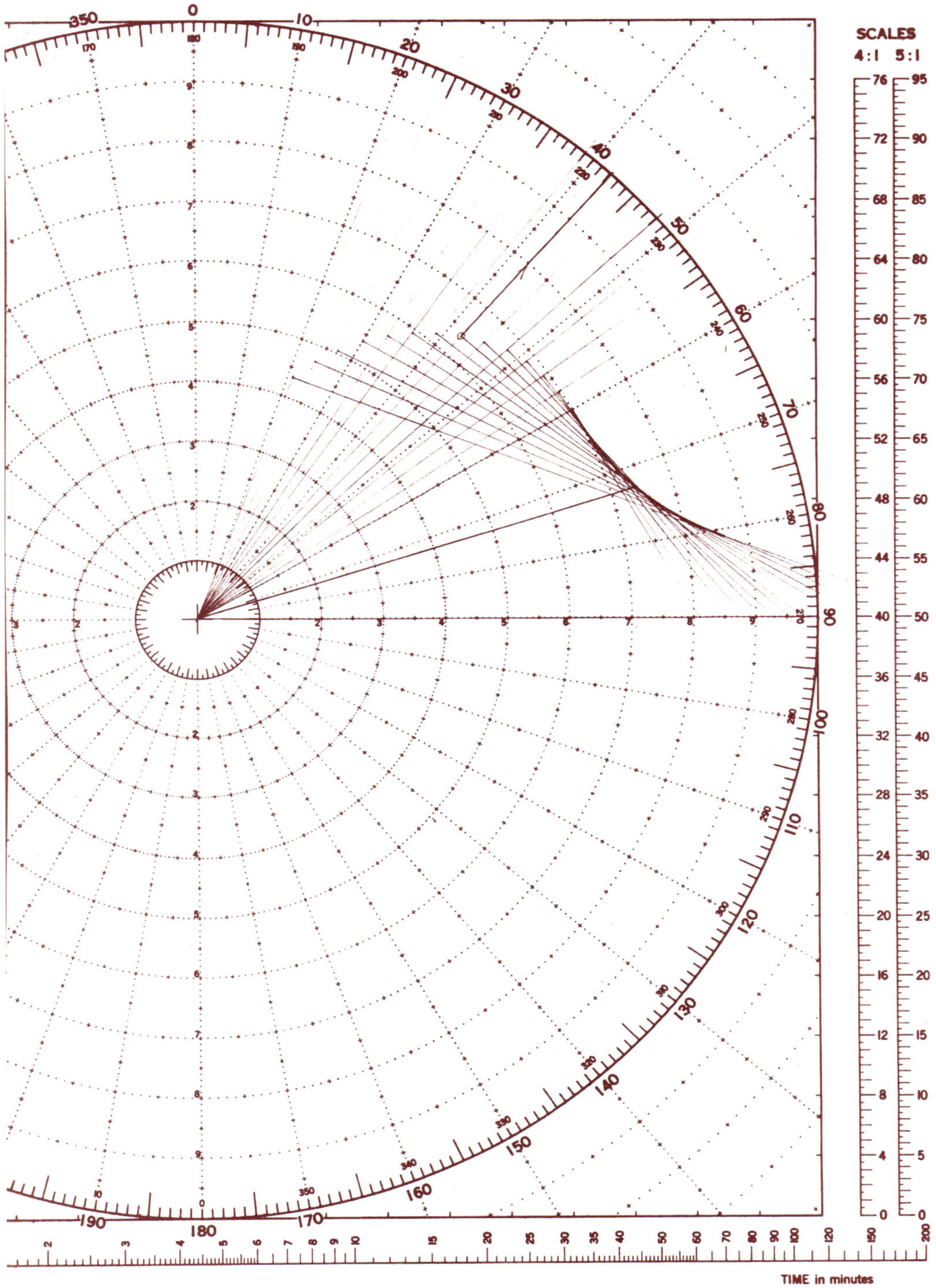
SCALES
4:1 5:1
TIME in minutes

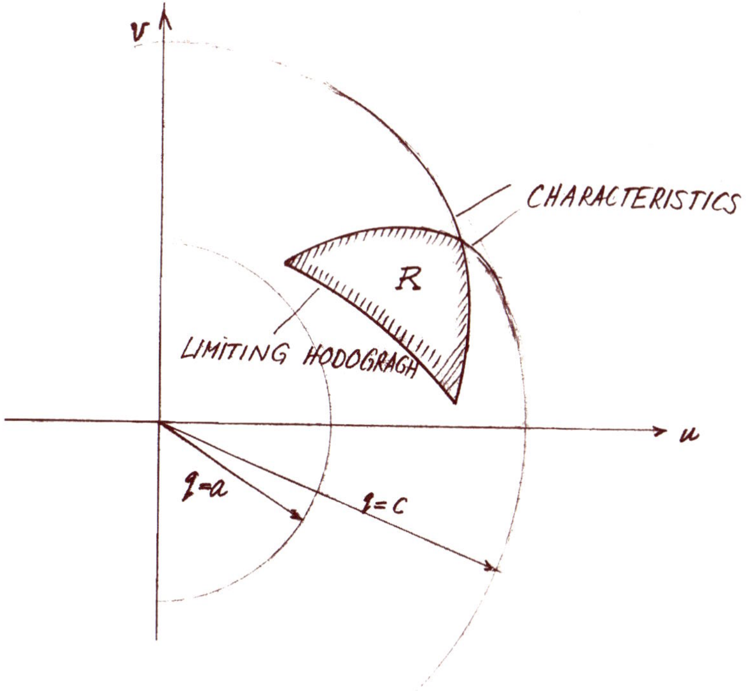

FIG. 5

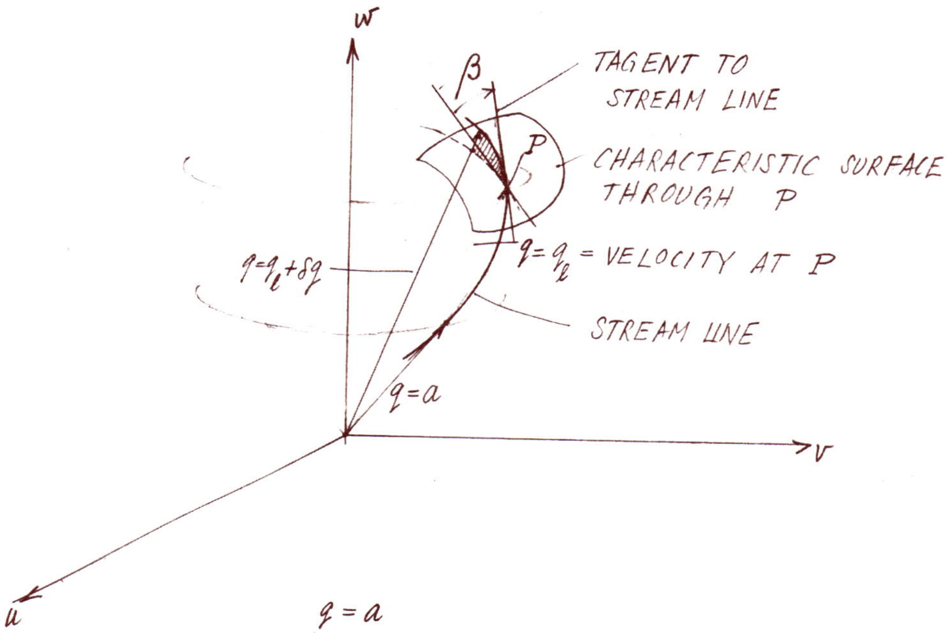

$q = a$

$q = q_l$

$q = C$

Fig 6

THE "LIMITING LINE" IN MIXED SUBSONIC AND SUPERSONIC FLOWS OF COMPRESSIBLE FLUIDS

Hsue-shen Tsien
California Institute of Technology

It is well-known that the vorticity for any fluid element is constant if the fluid is non-viscous and the change of states of the fluid is isentropic. When a solid body is placed in a uniform stream, the flow far ahead of the body is irrotational. Then if the flow is further assumed to be isentropic, the vorticity will be zero over the whole field of flow. In other words, the flow is irrotational. For such flow over a solid body, it is shown by T. Theodorsen (Ref. 1) that the solid body experiences no resistance. If the fluid has a small viscosity, its effect will be limited in the boundary layer over the solid body and the body will have a drag due to the skin friction. This type of essentially isentropic irrotational flow is generally observed for a stream-lined body placed in a uniform stream, if the velocity of the stream is kept below the so-called "critical speed".

At the "critical speed" or rather at a certain value of the ratio of the velocity of the undisturbed flow and the corresponding velocity of sound, shock waves appear. This phenomenon is called the "compressibility bubble", and the ratio of velocities. Along a shock wave, the change of states of the fluid is no more isentropic, although still adiabatic. This results in an increase in entropy of the fluid and generally introduces vorticity in an originally irrotational flow. The increase in entropy of the fluid is, of course, the consequence of changing part of the mechanical energy into heat energy. In other words, the part of fluid affected by the shock wave has much less mechanical energy. Therefore, with the appearance of shock waves, the wake of the stream line body is very much widened, and the drag increases drastically. Furthermore, the accompanied change in the pressure distribution over the body changes the aerodynamic moment acting on it and in case of an airfoil decreases the lift force.

of extremely large viscous effect. In other words, the boundary layer will be so thick as to occupy the main portion of the nozzle passage. To demonstrate this effect, let ~~us consider that~~ the length of the test section be L and the width of the square test section be b. Then the Reynolds number based upon the conditions in the test section is $Re = \frac{UL}{\nu}$ where U is the velocity in the test section. If, as a rough estimate, we take the thickness of the boundary layer to be zero at the beginning of the test section and equal to a value δ calculated by the well-known Blasius formula for a flat plate at the end of the test section, then

$$\delta = 3.65\, L \frac{1}{\sqrt{Re}} \tag{1}$$

Now if this boundary layer actually occupies half the tunnel width b/2, then

$$\delta = \frac{b}{2} = 3.65\, L \frac{1}{\sqrt{Re}} \tag{2}$$

On the other hand, the ratio of the mean free path l and the boundary layer thickness δ is known (Ref. 2) to be equal to

$$\frac{l}{\delta} = \frac{1.255\sqrt{\gamma}}{3.65} \frac{M}{\sqrt{Re}} \tag{3}$$

where γ is the ratio of specific heats and can be taken as 1.4, M is the Mach number in the test section. By combining (2) and (3), we have

$$\left(\frac{l}{\delta}\right)\left(\frac{L}{b}\right) = 0.0557\, M \tag{4}$$

This relation is shown in Fig. 1. Thus for a Mach number M equal to 2, and L/b = 2, the boundary layer will completely fill up the test section, if the mean free path is equal to 5.6% of the boundary layer thickness or 2.8% of the tunnel width. This means that the extremely strong viscous effect at low densities makes the ordinary concept of designing a wind tunnel totally inapplicable.

The ~~same fact~~ large viscous effects can be also demonstrated by calculating the ratio of frictional loss on the walls of the test section and the shock loss in the diffuser after the test section. Consider the diffuser to be a straight tube of approximately the same cross-sectional area as the test section, then the pressure loss due to friction Δp_1, is

(margin note: move 1 reference)

(margin note: main stream velocity at the center of the nozzle is supersonic. This is of course a new phenomenon in superaerodynamics tunnel not found in conventional supersonic wind tunnels)

The extremely thick boundary layer where the velocity increases from a small value near the wall to some supersonic velocity at the center of the nozzle, also gives rather subsonic velocities in a rather large portion of the nozzle. Since pressure disturbances downstream can be transmitted upstream in subsonic flows, the flow in the test section of a low pressure tunnel will be sensitive to changes in the diffuser even if the

-3-

Now if p is the static pressure in the test section, then the pressure by ideal isentropic compression in the diffuser is $p\left\{1+\frac{\gamma-1}{2}M^2\right\}^{\frac{\gamma}{\gamma-1}}$. If the actual pressure rise in the diffuser is estimated as that due to a normal shock without further recovery, then the actual pressure rise is $\left\{\frac{2\gamma}{\gamma+1}M^2-\frac{\gamma-1}{\gamma+1}\right\}p$.

$$\Delta p_1 = \frac{\textbf{Frictional Force}}{b^2}$$

Therefore the pressure loss due to shock Δp_2 is

$$= \frac{\rho U^2}{2} 4bL \cdot C_f \frac{1}{b^2}$$

Taking C_f to be Blasius value ~~again~~ or $C_f = \frac{1.328}{\sqrt{Re}}$, we have

$$\Delta p_1 = 2\rho U^2\left(\frac{L}{b}\right)\frac{1.328}{\sqrt{Re}} \tag{5}$$

~~Now the shock loss can be estimated as that due to a normal shock with no recovery after the shock. Then if p is the static pressure in the test section, the pressure loss due to shock Δp_2 can be calculated as the difference between the~~

Move above

$$\Delta p_2 = \left[\left\{1+\frac{\gamma-1}{2}M^2\right\}^{\frac{\gamma}{\gamma-1}} - \left\{\frac{2\gamma}{\gamma+1}M^2-\frac{\gamma-1}{\gamma+1}\right\}\right]p \tag{6}$$

By combining (5) and (6), the ratio of these two pressure losses is

$$\frac{\Delta p_1}{\Delta p_2} = \frac{2\gamma M^2\left(\frac{L}{b}\right)\frac{1.328}{\sqrt{Re}}}{\left[\left\{1+\frac{\gamma-1}{2}M^2\right\}^{\frac{\gamma}{\gamma-1}} - \left\{\frac{2\gamma}{\gamma+1}M^2-\frac{\gamma-1}{\gamma+1}\right\}\right]} \tag{7}$$

Introducing the mean free path ratio given by (3), we have

$$\frac{\Delta p_1}{\Delta p_2} = \left(\frac{L}{b}\right)\left(\frac{l}{\delta}\right)\frac{6.528\,\gamma M}{\left[\left\{1+\frac{\gamma-1}{2}M^2\right\}^{\frac{\gamma}{\gamma-1}} - \left\{\frac{2\gamma}{\gamma+1}M^2-\frac{\gamma-1}{\gamma+1}\right\}\right]} \tag{8}$$

This relation is plotted in Fig. 2. Therefore if the Mach number M is 2, and $L/b = 2$ as before, then when the ratio (l/δ) is 0.056, the ratio of frictional loss to shock loss is 0.628. Hence the frictional loss and the shock loss is of the same order of magnitude.

These large viscous effects are fully confirmed by the recent tests on the 1"×1" low pressure wind tunnel at the University of California.* The test nozzle (Fig 3) was designed for Mach number 4 without considering the viscous effect of the medium. During test, the static pressure on the wall at the exit of the nozzle is measured.

---

* Experimental work done under contract with the Office of Naval Research. The author is deeply indebted to Professors R. G. Folsom and E. D. Kane for permission to use their unpublished results.

this pressure is equal to 175 microns and 68 microns* for the two tests presented in Figs. 4 and 5. The apparent Mach numbers "M" is the Mach number calculated from the dynamic pressure measured by a Pitot tube by using the Rayleigh formula. Since there is the complication of large viscous effect in the Pitot tube ~~reading~~ as shown in the following section, this apparent Mach number is only qualitative and cannot be taken as the exact value. However it is apparent from Figs 4 + 5 that the boundary layer in the test section is indeed very thick, and fills up the whole space. This large boundary layer thickness makes the space available for the expansion of the central potential flow, if it exists, very small. Therefore the maximum Mach number reached at the centre of the nozzle is very much smaller than the designe Mach number of 4. At the lower pressure, the influence of slip at the wall is also evident. This has the tendency to make the flow ~~more~~ uniform. However the very low Mach number at the test section indicates again the strong viscous effect in converting much of the pressure energy into heat energy.

These elementary calculations and preliminary test results makes it clear that for the design of the nozzle and test section for a superaerodynamics wind tunnel, it is no longer possible to separate the compressibility effects and the viscous effect. In fact, the concept of boundary layer is also of doubtful value due to the extremely small Reynolds number encountered. Therefore to actually design such a nozzle to obtain the nearest approximation to the ideal uniform flow it will be necessary to use the exact Navier-Stokes equations instead of

* 1000 microns = 1 mm Hg, one atmosphere = 0.760×10^6 microns

the approximate boundary layer equations. Of course, it may be argued that for superaerodynamics, the Navier-Stokes equations for no more exact and additional corrections must be added (Ref. 2). However, recent investigations by R. Schamberg (Ref. 3) have shown that these additional corrections are small in case of slip-flows concerned here and will not essentially alter the flow pattern. Hence for a first approximation just like the non-viscous isentropic flow as a first approximation for ordinary supersonic nozzles, we can use the Navier-Stokes equations. The simplest case to be considered is certainly the axially symmetric nozzles. If x is the coordinate in the axial direction, r the coordinate in the radial direction and u and v are the corresponding velocity components, (Figure) the equations are:

$$\frac{\partial \rho u}{\partial x} + \frac{1}{r}\frac{\partial}{\partial r}(\rho r v) = 0 \tag{9}$$

$$\rho \frac{Du}{Dt} = -\frac{\partial p}{\partial x} + Grad(\tau)_x \tag{10}$$

$$\rho \frac{Dv}{Dt} = -\frac{\partial p}{\partial r} + Grad(\tau)_r \tag{11}$$

$$\rho \frac{D}{Dt}\left(\frac{u^2+v^2}{2} + c_p T\right) = \Phi - \left\{u\, Grad(\tau)_x + v\, Grad(\tau)_r\right\} + \frac{\partial}{\partial x}\left(\lambda \frac{\partial T}{\partial x}\right) + \frac{1}{r}\frac{\partial}{\partial r}\left(\lambda r \frac{\partial T}{\partial r}\right) \tag{12}$$

where $\frac{D}{Dt} = u\frac{\partial}{\partial x} + v\frac{\partial}{\partial r}$

Φ = dessipation function

τ = stresses tensor

(9), (10), (11) and (12) together with the equation of states

$$\frac{p}{\rho} = RT \tag{13}$$

then determines the five unknows u, v, p, ρ, and T. Of course, the actual process of making this calculation will be extremely tedious and some approximation method of solution may have to be developed. One possibility would be to adopt the Karman-Pqlhausen method for boundary layer to this case: We integrate the differential equations once with respect to r and thus only try to satisfy the equations "on the average" over the cross-section of the nozzle. The "distribution" of u, v over the cross-section will then be set in the form of a polynomical in r. Initial study in this procedure is already made by S. A. Schaaf (Ref. 4) at the suggestion of the author.

For ordinary supersonic diffuser, high efficiency of pressure recovery can be generally achieved by using a long diffuser. However, for superaerodynamics wind tunnel, due to the extremely large loss through friction, long diffusers are undesirable. In fact, the pressure loss can be reduced by using a shortest possible diffuser.

2. Flow Measurement

The quantities which determine the flow field are two out of the three variables p, ρ, T and the velocity components. The quantities p, ρ, T are related by the equation of states and therefore only two is necessary for the determination of all three. Generally for wind tunnel work, the quantities actually measured are p, ρ and q, the magnitude of the velocity.

For the measurement of pressure, a manometer is used. For ordinary pressure, one uses a fluid manometer filled with water, alcohol or mercury. However for the extremely low pressure encountered in the superaerodynamic flow, some other form of manometer is necessary. One of the most ~~used~~ successful type is the Pirani gauge. The conventional form of Pirani gauges has a pressure sensitivity of about 10^{-2} micron.* It utilizes the change of temperature of a wire heated with constant energy caused by a change in the pressure of the gas surrounding it. The temperature change is measured by the change in the resistance of the wire. The wire is located in a small chamber which is connected to the point of measurement by a hole, flush with the gas stream if static pressure is to be measured. The question of best design of the connecting tube for quick response is studied by S. A. Schaaf. (Ref. 5).

To measure the density ρ, the conventional method utilizes the difference in the velocity of light rays in mediums of different density. With different optical arrangements, we have the shadowgraph method, the schlieren method and the interferometer method. However, if the density of the medium is very low as the case of superaerodynamic flow, the sensitivity of these methods become extremely poor. For instance in case of schlieren method, the percentage change in illumination I by passing through a region of thickness

* 1000 micron = 1 m m. Hg., one atmosphere = 0.760×10^6 micron.

b is given by
$$\frac{\Delta I}{I} = k\frac{f}{e}\,0.000294\left(\frac{\rho}{\rho_0}\right)\left[\frac{b}{\rho}\frac{\Delta\rho}{\Delta n}\right] \tag{14}$$

where ρ_0 is the air density at 32°F and 1 atmosphere pressure, and $\Delta\rho/\Delta n$ is the density gradient normal to the light ray. f and e are the local length and the normal unobscured width of the light source image perpendicular to the knife edge. k is factor of order 1, determined by the particular optical path used. Therefore the sensitivity of the schlieren method decreases with the factor (ρ/ρ_0). Some imporvement can be made by altering the quantities f and e, but practical limitations and diffraction difficulties do not allow the increase of sensitivity to satisfactory values.

A new approach to this problem density meansurement is the method of absorption. It is found for instance, that oxygen at low pressures shows a strong absorption band at wave lengths around ~~1350~~ 1470 Å or ultra-violet light. The percentage absorption is proportional to the number of molecules that meet the light ray and is, therefore, proportional to the density of the gas. The measurement is then similar to that of the interferometer method where the density of gas is determined. A similar method is the utilization of the after-flow of nitrogen. These methods are now being studied by R. A. Evans (Ref. 6).

The conventional method for the measurement of velocity is through the use of dynamic pressure rise in a Pitot-tube. A straight forward application of this method is, however, difficult for rarefied gases. The formula used is, however, based upon the neglection of viscosity effects. But for rarefied gases the viscosity effect is of great importance as pointed out in the previous section. Then the dynamic pressure would be quite different than that given by the usual formula. To estimate this effect, let us consider the case of low Mach number so that compressibility effects can be neglected. Then as a first approximation, take the flow field around the Pitot-tube as that of a source of strength S in non-viscous flow of uniform velocity U. (~~Fig. 1~~). The "radius" of the tube a is

$$a = \sqrt{\frac{S}{\pi U}}$$

and the stagnation point is located at

$$R_s = \sqrt{\frac{\rho S}{4\pi U}} = \frac{\sqrt{3}}{2} a \tag{15}$$

The velocity introduced by the source is then

$$-U \frac{3}{4} \frac{a^2}{R^2}$$

By calculating the viscous stress from this approximate disturbance velocity, we have for flow along the axis

$$u\frac{\partial u}{\partial r} + \frac{1}{\rho}\frac{\partial p}{\partial r} = \nu U a^2 \frac{1}{2}\frac{1}{r^4} \tag{16}$$

Hence if p_0 is the stagnation pressure and p^0 the static pressure,

$$p_0 - p^0 = \frac{1}{2}\rho U^2 + \mu U a^2 \frac{1}{2}\int_\infty^{r_s} \frac{dr}{r^4} = \frac{1}{2}\rho U^2 - \frac{1}{2}\mu U \frac{a^2}{r_s^3}$$

Or

$$p_0 - p^0 = \frac{1}{2}\rho U^2 \left[1 - \frac{8}{9\sqrt{3}}\frac{\nu}{aU}\right] \tag{17}$$

For rarefied gases, the value of ν/aU or the reciprocal of the Reynold's number of the Pitot-tube could be of the order of unity. Then the dynamic pressure rise $p_0 - p^0$ is not the usual values $\frac{1}{2}\rho U^2$ but a value much less than that. ~~We shall be seriously in error if we use the ordinary formula to calculate the velocity U~~. In fact previous investigations by Baker (Ref. 7) and F. Homann (Ref. 8) indicate that the Reynolds number $\frac{a'U}{\nu}$ ~~must exceed~~ where a' is the radius of the mouth of tube, must ~~exceed~~ 30 in order for the to reach the usual dynamic pressure rise $\frac{1}{2}\rho U^2$.

When the velocity of flow is high, we have the added complication due to the shock. The conventional Rayleigh formula for Pitot tubes in supersonic flow is based upon the assumption of very thin shock wave ahead of the Pitot tube. Now the thickness of the shock is proportional to the mean free path. Hence in rarefied flows, the thickness of the shock will be so increased as to cause interference with flow in the neighborhood of the Pitot-tube. This together with the viscous effect mentioned in previous paragraph definitely show the inapplicability of the Rayleigh formula for supersonic velocity of rarefied gases.

Hot-Wire Anemometer

With the great complications in applying the conventional velocity measuring device to superaerodynamic flows, one is naturally led to the thought of other avenues of approach. One possibility is the use of hot-wire. If the wire diameter is of the order of 0.0001 inches, and if the pressure of the gas stream is approximately 100 microns, the ratio of the mean free path to the wire diameter will be approximately 180. Therefore the flow around the wire is definitely the free molecular flow (Ref. 2). We have thus a simple physical situation, which is an improvement over the rather uncertain circumstances of mixed dynamic and viscous effects for the measurement of velocity by Pitot tube. It thus seems worthwhile to explore this possibility by a trial calculation of the performance of such a hot-wire anemometer.

If θ is the inclination of the solid surface to a gas stream which has a macroscopic velocity U and a Maxwellian molecular velocity distribution, the translational energy of molecules $E_{i_{tr}}$ incident upon the unit area is

$$E_{i_{tr}} = \rho \frac{c}{2\sqrt{\pi}} \left\{ e^{-\left(\frac{U}{c}\right)^2 \sin^2\theta} (c^2 + \tfrac{1}{2}U^2) + \sqrt{\pi}\,\frac{U}{c}\sin\theta \left(\tfrac{5}{4}c^2 + \tfrac{1}{2}U^2\right)\left[1 + \mathrm{erf}\left(\frac{U}{c}\sin\theta\right)\right]\right\} \tag{18}$$

where $c^2 = 2RT$, T temperature of the gas stream and erf is the error function. Now let r be the radius of the hot-wire. Then the total energy E_i incident upon a unit length of the wire is the sum of translational energy and internal energy. If C_v is the specific heat at constant volume, this total energy per unit length of wire is

$$E_i = \rho \frac{c}{\sqrt{\pi}}\, r \left[\left\{ \tfrac{1}{2}U^2 + \left(\tfrac{1}{2}R + C_v\right)T \right\} \int_{-\frac{\pi}{2}}^{\frac{\pi}{2}} e^{-\left(\frac{U}{c}\right)^2 \sin^2\theta} d\theta \right.$$

$$\left. + \sqrt{\pi} \left\{ \tfrac{1}{2}U^2 + (R + C_v)T \right\} \int_{-\frac{\pi}{2}}^{\frac{\pi}{2}} \frac{U}{c}\sin\theta \left\{ 1 + \mathrm{erf}\left(\frac{U}{c}\sin\theta\right) \right\} d\theta \right] \tag{19}$$

(The functions F_1 and F_2 are tabulated in Table 1.)

The integrals in equation (19) can be expressed in terms of ~~tabulated~~ functions, (see Appendix) thus

$$E_i = \rho c r \left[\left\{ \tfrac{1}{2}U^2 + (\tfrac{1}{2}R + c_v)T \right\} F_1\left(\frac{U}{c}\right) + \left\{ \tfrac{1}{2}U^2 + (R + c_v)T \right\} F_2\left(\frac{U}{c}\right) \right] \quad (20)$$

where

$$F_1\left(\frac{U}{c}\right) = \frac{1}{\sqrt{\pi}} \int_{-\frac{\pi}{2}}^{\frac{\pi}{2}} e^{-\frac{U^2}{c^2}\sin^2\theta} \, d\theta = \sqrt{\pi}\, e^{-\frac{1}{2}\left(\frac{U}{c}\right)^2} I_0\left(\frac{1}{2}\frac{U^2}{c^2}\right) \quad (21)$$

$$F_2\left(\frac{U}{c}\right) = \int_{-\frac{\pi}{2}}^{\frac{\pi}{2}} \left(\frac{U}{c}\sin\theta\right) \left\{ 1 + \operatorname{erf}\left(\frac{U}{c}\sin\theta\right) \right\} d\theta = \sqrt{\pi} \left(\frac{U}{c}\right)^2 e^{-\frac{1}{2}\left(\frac{U}{c}\right)^2} \left[I_0\left(\frac{1}{2}\frac{U^2}{c^2}\right) - I_1\left(\frac{1}{2}\frac{U^2}{c^2}\right) \right] \quad (22)$$

The I_0 and I_1 are the modified Bessel functions of the first kind of orders zero and one respectively.

If T_w is the wall temperature, and α the accommodation coefficient, the difference between the energy E_i incident ~~upon~~ the surface and the energy E_r carried by the molecules re-emitted ~~from~~ the surface ~~is~~ given by

$$E_i - E_r = \alpha (E_i - E_w)$$

where E_w is the energy that would be carried away by the molecules if the re-emission were at the temperature T_w of the wire. Therefore

$$E_i - E_r = \alpha \rho c r \left[\left\{ \tfrac{1}{2}U^2 + (\tfrac{1}{2}R + c_v)(T - T_w) \right\} F_1\left(\frac{U}{c}\right) + \left\{ \tfrac{1}{2}U^2 + (R + c_v)(T - T_w) \right\} F_2\left(\frac{U}{c}\right) \right] \quad (23)$$

This difference of energy is ~~then~~ the net energy ~~input~~ to the ~~wire~~ per unit length of the ~~wire~~ by the air stream.

If i is the ~~electric~~ current heating ~~the~~ wire and $\omega\Omega$ is the resistance of the wire per unit length at the wire temperature, the heat input per unit length of wire by the heating current is

$i^2\Omega$. Heat is lost from the wire by radiation. If σ is Stefan-Boltzmann constant and ε is the emissivity of the wire surface, the radiation heat loss per unit length is $2\pi r \varepsilon \sigma T_w^4$. Therefore if the wire has reached a steady condition, the heat balance requires

$$\alpha\rho\sqrt{2RT}\, r\left[F_1\left(\frac{U}{c}\right)\left\{\frac{1}{2}U^2+\left(\frac{1}{2}R+c_v\right)(T-T_w)\right\}+F_2\left(\frac{U}{c}\right)\left\{\frac{1}{2}U^2+(R+c_v)(T-T_w)\right\}\right]$$

$$+\, i^2\Omega = 2\pi r \varepsilon \sigma T_w^4 \qquad (24)$$

This equation can be put into somewhat simpler form by using the relation that

$$R = c_p - c_v = c_v(\gamma-1) \qquad (25)$$

Furthermore if we take T_0 is ~~the~~ reference temperature at which the resistance Ω is Ω_0, and the corresponding temperature coefficient of the resistance is β. ~~Then~~ the resistance Ω can be expressed as

$$\Omega = \Omega_0\left[1+\beta(T_w-T_0)\right] \qquad (26)$$

Now let

$$\lambda = \Omega/\Omega_0 \qquad (27)$$

then from equation (26)

$$\frac{T_w}{T_0} = \frac{\lambda-1}{\beta T_0} + 1 \qquad (28)$$

Now introduce ρ_0 as the reference density, and i_0 as the reference heating current, then equation (24) can be written as

$$\left\{1+\frac{\lambda-1}{\beta T_0}\right\}^4 = \left[\frac{\alpha\rho_0 (RT_0)^{3/2}}{\pi\sqrt{2}\,\varepsilon\sigma T_0^4}\right]\left(\frac{\rho}{\rho_0}\right)\left(\frac{T}{T_0}\right)^{3/2}\left[F_1\left(\frac{U}{c}\right)\left\{\left(\frac{U}{c}\right)^2+\frac{1}{2}\frac{\gamma+1}{\gamma-1}\left(1-\frac{1+\frac{\lambda-1}{\beta T_0}}{T/T_0}\right)\right\}\right.$$

$$\left.+\,F_2\left(\frac{U}{c}\right)\left\{\left(\frac{U}{c}\right)^2+\frac{\gamma}{\gamma-1}\left(1-\frac{1+\frac{\lambda-1}{\beta T_0}}{T/T_0}\right)\right\}\right]+\left(\frac{i_0^2\Omega_0}{2\pi r\varepsilon\sigma T_0^4}\right)\lambda\left(\frac{i}{i_0}\right)^2 \qquad (29)$$

The particular values of the reference temperature T_0, the reference density ρ_0 and the reference current i_0 are not yet fixed. We fix these quantities now by ~~requiring~~ that

$$\beta T_0 = 1 \tag{30}$$

$$\frac{\alpha p_0 (RT_0)^{3/2}}{\pi\sqrt{2}\,\varepsilon\sigma T_0^4} = 1 \tag{31}$$

and

$$\frac{i_0^2 \Omega_0}{2\pi r \varepsilon \sigma T_0^4} = 1 \tag{32}$$

Then equation (29) simplifies into

$$\lambda^4 = \left(\frac{p}{p_0}\right)\left(\frac{T}{T_0}\right)^{1/2}\Big[F_1\left(\frac{u}{c}\right)\left\{\left(\frac{u}{c}\right)^2 + \frac{1}{2}\frac{\gamma+1}{\gamma-1}\left(1 - \frac{\lambda}{T/T_0}\right)\right\}$$

$$+ F_2\left(\frac{u}{c}\right)\left\{\left(\frac{u}{c}\right)^2 + \frac{\gamma}{\gamma-1}\left(1 - \frac{\lambda}{T/T_0}\right)\right\}\Big] + \lambda\cdot\left(\frac{i}{i_0}\right)^2 \tag{33}$$

This is then the performance equation of the hot-wire in free molecular flow.

Now let ~~us~~ investigate in greater detail the case of a bright platinum wire. To satisfy equation (30),

$$T_0 = 492^\circ R.$$

The value for ε and α can be ~~taken to be~~ 0.08 and 0.90 ~~respectively~~. Then equation (31) gives the corresponding pressure p_0 for ρ_0 and T_0 as

$$p_0 = \frac{\pi\sqrt{2}\;\varepsilon\sigma T_0^4}{\alpha\sqrt{RT_0}} = 3.37 \text{ microns.}$$

Let the radius of the wire be 0.0001 inch. Then equation (32) gives the reference heating current i_0 as

$$i_0 = \sqrt{\frac{2\pi r \varepsilon \sigma T_0^4}{\omega \Omega_0}} = 0.274 \text{ milliampere.}$$

where the resistivity of the platinum is taken as 10.96×10^{-6} ohms-cms. Therefore the order of magnitude of the different quantities is entirely satisfactory.

If the wire is used with a constant heating current, then equation (33) can be used to calculate the relation between the resistance ratio λ and the velocity ratio $\left(\frac{u}{c}\right)$ at constant air steam density and temperature. This is done for $p/p_0 = 1$, $T/T_0 = 1$ and $i/i_0 = 1$ * and the result is given in Fig. 6. It is seen that the sensitivity of the instrument is good. Of course, the behavior of the hot-wire anemometer will be actually determined by calibration for any experiment. However the present analysis seems to indicate the feasibility of such an instrument, and further research is definitely desirable. for measurements in rarefied gases

4. Parameters of Flow

Since the performance of the wire is strongly influenced by the accommodation coefficient α as shown by equation (28), it will be necessary to find materials which can hold this coefficient constant for a considerable period of time so that no frequent calibration is required.

---

* The author is indebted to Mr. L. Mack for the numerical computations

gas velocity U, gas temperature T and gas density ρ and the resistance r will be obtained by direct calibration. But nevertheless the sensitivity of the instrument will be of the same order as shown above. Therefore these preliminary investigations seem to indicate the desirability of further research on the hot wire anemometer for rarefied gases.

4. Parameters of Flow

The two parameters that are directly connected with the flow field are the Reynold's number Re, defined as

$$Re = \frac{UL}{\nu^{\circ}}$$

where ν° is the kinematic viscosity, L the typical linear dimension of the body; and the Mach number M° of the free stream. This is true even for slip flows and free molecule flows due to the fact that the ratio of mean free path to the typical dimension can be also expressed in terms of the Reynold's number and the Mach number.

However, as the pressure or density is reduced, the solid boundary of the flow enters actively into the flow conditions by requiring not only that the microscopic stream velocity be tangential to the surface but that the interaction of the molecules and the wall be considered and that the radiation of energy to and from the wall be taken into account. The interaction of the molecules with the wall is so far expressed through the fraction s of molecules that are diffusely re-emitted from the wall and the accommodation coefficient α. We have reasons to believe that both s and α are functions of the temperature of the wall and the molecular velocity distribution. Therefore the interaction of the molecules with the wall is the same only if the wall temperature, the gas temperature and the Mach number of the gas above the wall is the same. These considerations seem to indicate then that for the model test to be similar to the prototype, the model must be made of same surface material as the prototype, and the fluid must be the same, and furthermore the following parameters must be the same:

(1) Reynolds number Re

(2) Mach number M°

(3) free stream temperature, T°

The radiation heat loss from the surface is equal to $\varepsilon\sigma T_w^4$ per unit area. However, if the model is surrounded by the walls of the test chamber, there is also an heat input due to radiation from walls of the test section to the model. Let us call this quantity q_c. Then the net heat loss per unit area of the surface of the model is $\varepsilon\sigma T_w^4 - q_c$. This quantity can be rendered non-dimensional by dividing it by $\rho^0 U (c_p T^0)$. Call this new parameter Λ_m, then

$$\Lambda_m = \left[\frac{\varepsilon\sigma T_w^4 - q_c}{\rho^0 U (c_p T^0)}\right]_m \qquad (\sout{25}\ 34)$$

For the prototype, the heat from the walls of the test chamber is absent but there may be solar radiation and the radiation from the earth and surrounding atmosphere. (Ref. 9). Denoting this amount by q, then the parameter Λ for the prototype is

$$\Lambda = \frac{\varepsilon\sigma T_w^4 - q}{\rho^0 U (c_p T^0)} \qquad (\sout{26}\ 35)$$

In order for the flow to be same also with respect to the radiation heat transfer,

$$\Lambda = \Lambda_m \qquad (\sout{27}\ 36)$$

Because of the previous conditions on the Reynolds number and free stream temperature, (~~27~~ 36) is the same as

$$\frac{\varepsilon\sigma T_w^4 - q}{\varepsilon\sigma T_w^4 - q_c} = \frac{L_m}{L} \qquad (\sout{28}\ 37)$$

where L_m is the typical linear dimension of the model and L is the typical linear dimension of the prototype. This means that the wall temperature of the test chamber must be so controlled that q_c satisfies (~~28~~ 37).

This set of rather strict similarly rules for model testing in super-aerodynamic flows is certainly difficult to satisfy. In what way the rules can be relaxed is the problem of future research.

THE "LIMITING LINE" IN MIXED SUBSONIC AND SUPERSONIC FLOW OF COMPRESSIBLE FLUIDS

Hsue-shen Tsien
California Institute of Technology

It is well-known that the vorticity for any fluid element is constant if the fluid is non-viscous and the change of state of the fluid is isentropic. When a solid body is placed in a uniform stream, the flow far ahead of the body is irrotational. Then if the flow is further assumed to be isentropic, the vorticity will be zero over the whole field of flow. In other words, the flow is irrotational. For such flow over a solid body, it is shown by T. Theodorsen (Ref. 1) that the solid body experiences no resistance. If the fluid has a small viscosity, its effect will be limited in the boundary layer over the solid body and the body will have a drag due to the skin friction. This type of essentially isentropic irrotational flow is generally observed for a stream-lined body placed in a uniform stream, if the velocity of the stream is kept below the so-called "critical speed".

At the "critical speed" or rather at a certain value of the ratio of the velocity the undisturbed flow and the corresponding velocity of sound, shock waves appear. This phenomenon is called the "compressibility burble". Along a shock wave, the change of state of the fluid is no more isentropic, although still adiabatic. This results in an increase in entropy of the fluid and generally introduces vorticity in an originally irrotational flow. The increase in entropy of the fluid is, of course, the consequence of changing part of the mechanical energy into heat energy. In other words, the part of fluid affected by the shock wave has much less mechanical energy. Therefore, with the appearance of shock waves, the wake of the stream line body is very much widened, and the drag increases drastically. Furthermore, the accompanied change in the pressure distribution over the body changes the aerodynamic moment acting on it and in case of an airfoil decreases the lift force.

All these consequences of the breakdown of isentropic irrotational flow are generally undesirable in applied aerodynamics. Its occurrence should be delayed as much as possible by modifying the shape or contour of the body. However, such endeavor will be very much facilitated if the cause or the criterion for the breakdown can be found first.

Criterion for the Breakdown of Isentropic Irrotational Flow

G. I. Taylor and C. F. Sharman (Ref. 2) calculated the successive approximations to the flow around an airfoil by means of an electrolyte tank. They found that when the maximum velocity in the flow reaches the local velocity of sound, the convergence of the successive steps seems to break down. This fact led to the identification of critical speed or critical Mach number with the Mach number of the undisturbed flow for which the local velocity at some point reaches the local velocity of sound. However, there is no mathematical proof for the coincidence of the critical Mach number so defined and the breakdown of isentropic irrotational flow. Furthermore, such a definition for critical Mach number implies that a transition from a velocity less than that of sound, or subsonic velocity, to a velocity greater than that of sound, or supersonic velocity, does not occur in isentropic irrotational flow. On the other hand, Taylor (Ref. 3) and others found solutions for which such a transition occurs. Furthermore, A. M. Binnie and S. G. Hooker (Ref. 4) have shown that at least for the case of spiral flow the method of successive approximation is a convergent one even for supersonic velocities. With these facts in mind, one can conclude that the identification of critical speed with local supersonic velocity cannot be correct.

Taylor's investigation on the spiral flow (Ref. 3) indicates that there is a line in the flow field where the maximum velocity is reached and beyond which the flow cannot continue. W. Tollmien in a subsequent paper (Ref. 5) called such lines as the "limiting lines". The velocity at the limiting line is never subsonic. However, the true characteristics of such limiting lines and their significance were not investigated by Tollmien at that time. Recently, F. Ringleb (Ref. 6) obtained another particular solution of isentropic irrotational flow in which the maximum velocity reached is approximately twice the local sound velocity. For this flow also, a limiting line appeared beyond which the flow cannot continue. Furthermore, he found the singular character of the limiting line, such as the infinite acceleration and infimite pressure gradient. Th. von Karman (Ref. 7) demonstrated this fact for the general two-dimensional flow. He also suggested that the limiting line is the envelope of the Mach waves (Fig. 1) and thus can only occur in supersonic region. He also took its appearance as the criterion for breakdown of isentropic irrotational flow. This general two-dimensional theory was established later by both Ringleb (Ref. 8) and Tollmien (Ref. 9). Tollmien corrected some mistakes in Ringleb's paper and in addition, showed that the flow definitely cannot continue beyond the limiting line. The later fact introduced a "forbidden region" in the flow bounded by the limiting line. This physical absurdity can only be avoided by relaxing the condition of irrotationality. But as stated previously, for non-viscous fluids, the transition from a flow without vorticity to that with vorticity can only be accomplished by shock waves, which at the same time also cause an increase in the entropy.

However, before one can conclude that the appearance of limiting line, or the envelope of Mach waves, is the general condition for breakdown of isentropic irrotational flow, one must prove that the singular behavior of limiting lines are general and not limited to two-dimensional flow. This is the purpose of the present paper. First the property of limiting line in axially symmetric flow will be investigated in detail. Then the general three-dimensional problem will be sketched. These investigations confirm the results of Ringleb, von Karman and Tollmien for these more general cases.

Therefore, by considering only the steady flow of non-viscous fluids, the criterion for breakdown of isentropic irrotational flow is the appearance of limiting line. However, for the actual motion of a solid body, the flow is neither steady nor non-viscous. Small disturbances always occur and almost all real fluids have appreciable viscosity. The small disturbances in the flow introduce the question of stability. In other words, the solution found for isentropic irrotational flow may be unstable even before the appearance of limiting line, and tends to transform itself to a rotational flow involving shock waves at the slightest disturbance. If this is the case, the criterion concerns not the limiting line, but the stability limit. This problem has yet to be solved.

The effect of viscosity will be limited to the boundary layer if the pressure along the surface in the flow direction never increases too rapidly. Then outside the boundary layer the flow is isentropic and irrotational. If the gradient of pressure is too large, the boundary layer will separate from the surface. However, at low velocities such separation only widens the wake of the body and changes the pressure distribution over the body. But if the boundary layer separates at a point where the velocity outside the boundary layer is supersonic, additional effects may appear. The flow outside the boundary layer in this case can be regarded approximately as that of a solid body not of original contour but of a new contour including the "dead water" region created by the separation. It is then immediately clear that the ideal isentropic irrotational flow around this new contour may have a limiting line. Hence, the actual flow then must involve shock waves. In other words, the separation of boundary layer in supersonic region may induce a shock wave and thus extend its influence far beyond the region of separation. Furthermore, the steep adverse pressure gradient across a shock wave may accent the separation. This interaction between the separation and the shock wave is frequently observed in experiments.

The above considerations indicate the possibility of the breakdown of isentropic irrotational flow outside the boundary layer even before the appearance of limiting line. Therefore, the Mach number of the undisturbed flow at which the limiting line appears may be called as the "upper critical Mach number." On the other hand, since shock wave can only occur in supersonic flow, the Mach number of the undisturbed flow at which local velocity reaches the velocity of sound may be called as the "lower critical Mach number". The actual critical Mach number for the appearance of shock waves and the compressibility burble must lie between these two limits. By carefully designing the contour of the body to avoid the crowding together of Mach waves to form an envelope and to eliminate adverse pressure gradient along the surface of the body, the compressibility burble can be delayed.

Axially Symmetric Flow

The solution of the exact differential equations for an axially symmetric isentropic irrotational flow was first given by F. Frankle (Ref. 10)

The method is developed independently by C. Ferrari (Ref. 11). Their method applies particularly to the case of supersonic flow over a body of revolution with pointed nose. In this case, the flow at the nose can be approximated by the well-known solution for a cone. From this solution, the differential equation is solved step by step using the net of characteristics which are real for supersonic velocities. In the following investigation, the chief concern is not the solution of the partial differential equation but rather the occurrence and the properties of the limiting line in an isentropic irrotational flow. The general plan of attack is that of Tollmein (Ref. 9). However, here the calculation is based on the Legendre transformation of velocity potential instead of the stream function.

If q is the magnitude of the velocity, a the corresponding velocity of sound assuming isentropic process, p the pressure and ρ the density of fluid, the Bernoulli equation gives

$$\frac{\rho}{\rho_0} = \left(1 - \frac{\gamma-1}{2}\frac{q^2}{a_0^2}\right)^{\frac{1}{\gamma-1}} = \left(1 + \frac{\gamma-1}{2}\frac{q^2}{a^2}\right)^{-\frac{1}{\gamma-1}} \tag{1}$$

$$\frac{a^2}{a_0^2} = 1 - \frac{\gamma-1}{2}\frac{q^2}{a_0^2} = \left(1 + \frac{\gamma-1}{2}\frac{q^2}{a^2}\right)^{-1} \tag{2}$$

$$\frac{p}{p_0} = \left(1 - \frac{\gamma-1}{2}\frac{q^2}{a_0^2}\right)^{\frac{\gamma}{\gamma-1}} = \left(1 + \frac{\gamma-1}{2}\frac{q^2}{a^2}\right)^{-\frac{\gamma}{\gamma-1}} \tag{3}$$

In these equations, the subscript o denotes quantities corresponding to $q=0$ and γ is the ratio of specific heats of the fluid. Let the axis of symmetry be the x - axis, the distance normal to x - axis be denoted by y, and the velocity components along these two directions be denoted by u and v respectively (Fig. 2). The x - y plane is, therefore, a meridian plane. Then the kinematical relations of the flow are given by the vorticity equation

$$v_x - u_y = 0^{*} \tag{4}$$

* Throughout this paper, partial derivatives are denoted by subscripts. Thus $v_x \equiv \frac{\partial v}{\partial x}$, $u_y \equiv \frac{\partial u}{\partial y}$.

and the continuity equation

$$\frac{\partial}{\partial x}\left(y \frac{\rho}{\rho_0} u\right) + \frac{\partial}{\partial y}\left(y \frac{\rho}{\rho_0} v\right) = 0 \tag{5}$$

Eqs. (1) (2) (3) (4) and (5) together with the relation $q^2 = u^2 + v^2$ specify the flow completely.

To simplify the problem, a velocity potential φ defined as follows is introduced:

$$u = \varphi_x, \quad v = \varphi_y \tag{6}$$

Then Eq. (4) is identically satisfied and Eq. (5) together with Eqs. (1) and (2) gives the equation for φ.

$$\left(1 - \frac{u^2}{a^2}\right)\varphi_{xx} - 2\frac{uv}{a^2}\varphi_{xy} + \left(1 - \frac{v^2}{a^2}\right)\varphi_{yy} + \frac{v}{y} = 0 \tag{7}$$

The characteristics of this differential equation, to be called the characteristics in the physical plane is given by $g(x,y) = 0$, where $g(x,y)$ is determined by the following equation

$$\left(1 - \frac{u^2}{a^2}\right) g_x^2 - 2\frac{uv}{a^2} g_x g_y + \left(1 - \frac{v^2}{a^2}\right) g_y{}^2 = 0 \tag{8}$$

It can be easily seen from this equation that g is real only when $q > a$. Therefore, the characteristics are real only in supersonic regions of the flow.

The meaning of characteristics in the plysical plane is immediately clear if one calculates the relation between the slope of a characteristic and the slope of a stream line in the meridian or xy plane. By the definition of the function <u>g(x,y)</u>, the value of <u>g</u> is zero, or constant, along a characteristic. Therefore, by writing a quantity evaluated at a certain constant value of a parameter with that parameter as a subscript, the slope of characteristic in physical plane is

$$\left(\frac{dy}{dx}\right)_g = - \frac{g_x}{g_y} \tag{9}$$

Along a stream line, the stream function ψ defined by following equations is constant:

$$\psi_y = y\frac{\rho}{\rho_0}u, \quad \psi_x = -y\frac{\rho}{\rho_0}v \tag{10}$$

Therefore, the slope of a stream line is

$$\left(\frac{dy}{dx}\right)_\psi = \frac{v}{u} \tag{11}$$

Eqs. (8), (9) and (11) give

$$\left(\frac{dy}{dx}\right)_q = \frac{-\frac{uv}{a^2} \pm \sqrt{\frac{q^2}{a^2}-1}}{1-\frac{u^2}{a^2}} = \left\{\left(\frac{dy}{dx}\right)_\psi \mp \tan\beta\right\} \div \left\{1 \pm \left(\frac{dy}{dx}\right)_\psi \tan\beta\right\} \tag{12}$$

where β is the Mach angle given by $\beta = \sin^{-1}\frac{a}{q}$. Therefore, Eq. (12) shows that the characteristics in physical plane are inclined to the stream lines by an angle equal to the Mach angle. Such lines are the wave fronts of infinitesimal disturbances and are called Mach waves. In ther words, characteristics in physical planes are the Mach waves in that plane. There are two families of Mach waves inclined symmetrically with respect to each stream line.

If to each pair of values of u and v, there is <u>one</u> pair of values of x, y, then x and y can be considered as functions of u, v. In other words, instead of taking x and y as independent variables, u, v can be used as independent variable. The plane with u, and v as coordinates is called the "hodograph plane." An equation in hodograph plane corresp nding to Eq. (7) can be obtained by means of Legendre's transformation. By writing

$$\chi = ux + vy - \varphi \tag{13}$$

it is seen that

$$\chi_u = x, \quad \chi_v = y \tag{14}$$

Then Eq. (7) can be written as

$$\left(1-\frac{u^2}{a^2}\right)\chi_{vv} + 2\frac{uv}{a^2}\chi_{uv} + \left(1-\frac{v^2}{a^2}\right)\chi_{uu} + \frac{v}{\chi_v}\left(\chi_{uu}\chi_{vv} - \chi_{uv}^2\right) = 0 \tag{15}$$

The characteristics of Eq. (15) are given by $f(u,v)=0$ where f is the solution of following differential equation

$$\left\{\left(1-\frac{u^2}{a^2}\right)+\frac{v}{\chi_v}\chi_{uu}\right\}f_v^2+2\left(\frac{uv}{a^2}-\frac{v}{\chi_v}\chi_{uv}\right)f_u f_v + \left\{\left(1-\frac{v^2}{a^2}\right)+\frac{v}{\chi_v}\chi_{vv}\right\}f_u^2=0 \tag{16}$$

Eq. (16) shows that the characteristics in hodograph plane depends upon the values of the derivatives of χ which must be obtained from Eq. (15). In other words, the characteristics in hodograph plane change with the flow and are not a constant set of curves as those in two-dimensional problems.

To obtain the relation between the characteristics in physical plane and those in hodograph plane, it is noticed that Eq. (9) can be rewritten as

$$(dy)_g:(dx)_g=-g_x:g_y \tag{17}$$

Then Eq. (8) is equivalent to

$$\left(1-\frac{u^2}{a^2}\right)(dy)_g^2+2\frac{uv}{a^2}(dy)_g(dx)_g+\left(1-\frac{v^2}{a^2}\right)(dx)_g^2=0 \tag{18}$$

However, in general, Eq. (14) gives the following relation between the differentials of x and y and those of u and v:

$$\begin{aligned} dx &= \chi_{uu}\,du+\chi_{uv}\,dv \\ dy &= \chi_{uv}\,du+\chi_{vv}\,dv \end{aligned} \tag{19}$$

By means of these relations, Eq. (18) can be transformed into an equation for $(du)_g$ and $(dv)_g$. This transformed equation can be simplified by using Eq. (15), the final relation is

$$\left(\chi_{uu}\chi_{vv}-\chi_{uv}^2\right)\left[\left\{\left(1-\frac{u^2}{a^2}\right)+\frac{v}{\chi_v}\chi_{uu}\right\}(du)_g^2-2\left(\frac{uv}{a^2}-\frac{v}{\chi_v}\chi_{uv}\right)(du)_g(dv)_g + \left\{\left(1-\frac{v^2}{a^2}\right)+\frac{v}{\chi_v}\chi_{vv}\right\}(dv)_g^2\right]=0 \tag{20}$$

Therefore, if the first factor of Eq. (20) is not zero, the variations $(du)_g$ and $(dv)_g$ along a characteristic in physical plane must satisfy the relation

$$\left\{\left(1-\frac{u^2}{a^2}\right)+\frac{v}{\chi_v}\chi_{uu}\right\}(du)_g^2-2\left\{\frac{uv}{a^2}-\frac{v}{\chi_v}\chi_{uv}\right\}(du)_g(dv)_g+\left\{\left(1-\frac{v^2}{a^2}\right)+\frac{v}{\chi_v}\chi_{vv}\right\}(dv)_g^2=0 \tag{21}$$

This is the same relation for the variations $(du)_f$ and $(dv)_f$ along a characteristics in hodograph plane as can be seen from Eq. (16) and the following relation obtained from the definition of

$$(dv)_f : (du)_f = -f_u : f_v \tag{22}$$

The transformed characteristics of physical plane and the characteristics of hodograph plane themselves satisfy then the same first order differential equation. Therefore, these two types of curves are the same. In other words, the characteristics of hodograph plane are the representation of Mach waves in u, v plane.

The Limiting Line

Eq. (20) shows that if

$$\chi_{uu}\chi_{vv}-\chi_{uv}^2=0 \tag{23}$$

then the transformed differential equation for characteristics of physical plane, or Mach waves is satisfied. Therefore, if there is a line in the hodograph plane, along which the values of the derivatives of χ are such that Eq. (23) is true, then this line when transferred to the physical plane will have its slope equal to that of one family of Mach waves. Such lines are called as limiting hodograph in u-v plane and limiting line in physical plane. Since Mach waves only occur in the supersonic regions, it is then evident that the limiting line must appear in these regions. The significance of the adjective "limiting" will be made clear as other properties of such lines are investigated.

Now the question arises: Can the limiting hodograph be the characteristic in u-v plane? Along a limiting hodograph, Eq. (23) gives

$$\left(\frac{dv}{du}\right)_l = -\frac{\chi_{uuu}\chi_{vv} - 2\chi_{uv}\chi_{uuv} + \chi_{uu}\chi_{uvv}}{\chi_{uuv}\chi_{vv} - 2\chi_{uv}\chi_{uvv} + \chi_{uu}\chi_{vvv}} \tag{24}$$

where the subscript l denotes the value along a limiting hodograph. Now the general differential equation for χ, Eq. (15), is true for the whole u-v plane, therefore, the equation is still true by differentiating it with respect to u and v. The results can be simplified by using Eq. (15) itself and Eq. (23). Then at the limiting hodograph,

$$\left[\left(1-\frac{v^2}{a^2}\right)+\frac{v}{\chi_v}\chi_{vv}\right]\chi_{uuu} + 2\left[\frac{uv}{a^2} - \frac{v}{\chi_v}\chi_{uv}\right]\chi_{uuv} + \left[\left(1-\frac{u^2}{a^2}\right)+\frac{v}{\chi_v}\chi_{uu}\right]\chi_{uvv}$$
$$= (\gamma+1)\frac{u}{a^2}\chi_{vv} - 2\frac{v}{a^2}\chi_{uv} + (\gamma-1)\frac{u}{a^2}\chi_{uv} \tag{25a}$$

$$\left[\left(1-\frac{v^2}{a^2}\right)+\frac{v}{\chi_v}\chi_{vv}\right]\chi_{uuv} + 2\left[\frac{uv}{a^2} - \frac{v}{\chi_v}\chi_{uv}\right]\chi_{uvv} + \left[\left(1-\frac{u^2}{a^2}\right)+\frac{v}{\chi_v}\chi_{uu}\right]\chi_{vvv}$$
$$= (\gamma+1)\frac{u}{a^2}\chi_{vv} - 2\frac{u}{a^2}\chi_{uv} + (\gamma+1)\frac{v}{a^2}\chi_{uu} \tag{25b}$$

Eqs. (24), (25a) and (25b) are the only available equations involving no higher derivative than the third. On the other hand, the slope of a characteristic in hodograph plane can be calculated by Eq. (22),

$$\left(\frac{dv}{du}\right)_f = -\frac{f_u}{f_v} \tag{26}$$

This equation together with Eq. (16) gives

$$\left\{\left(1-\frac{v^2}{a^2}\right)+\frac{v}{\chi_v}\chi_{vv}\right\}\left(\frac{dv}{du}\right)_f^2 - 2\left\{\frac{uv}{a^2} - \frac{v}{\chi_v}\chi_{uv}\right\}\left(\frac{dv}{du}\right)_f + \left\{\left(1-\frac{u^2}{a^2}\right)+\frac{v}{\chi_v}\chi_{uu}\right\} = 0 \tag{27}$$

Therefore, if the limiting hodograph is a characteristic, then $\left(\frac{dv}{du}\right)_l$ must satisfy Eq. (27). However, a simple calculation shows that it is not even possible to obtain a relation between $\left(\frac{dv}{du}\right)_l$ and other quantities not involving the third order derivatives of χ. Hence, $\left(\frac{dv}{du}\right)_l$ does not satisfy Eq. (27). In other words, the limiting hodograph is not a characteristic. Transferred to physical plane, this means that the limiting line is not a Mach wave. But as shown in previous paragraphs, the limiting line is everywhere tangent to one family of Mach waves. Consequently, the limiting line must be the envelope of a family of Mach waves. This property of limiting line can be taken as its physical definition.

Limiting Hodograph and the Stream Lines

At the limiting hodograph both Eqs. (15) and (23) hold, by eliminating one of the second order derivatives, say χ_{uu}, the following relation is obtained

$$(\chi_{vv})_l = \frac{-\frac{uv}{a^2} \pm \sqrt{\frac{q^2}{a^2}-1}}{1-\frac{u^2}{a^2}} (\chi_{uv})_l \tag{28}$$

The sign before the radical in Eq. (28) can be either positive or negative but not both. This relation will be used presently to show that the stream lines and one family of characteristics are tangent in u-v plane.

From Eq. (10), the differential of stream function can be calculated as

$$d\psi = -y\frac{\rho}{\rho_0} v\,dx + y\frac{\rho}{\rho_0} u\,dy \tag{29}$$

In this equation, y can be replaced by χ_v according to Eq. (14) and the differentials dx and dy replaced by the differential du and dv according to Eq. (19). Then

$$d\psi = \chi_v \frac{\rho}{\rho_0}\left[(-v\chi_{uu}+u\chi_{uv})\,du + (-v\chi_{uv}+u\chi_{vv})\,dv\right] \tag{30}$$

Along a stream line, $d\psi = 0$, therefore the slope of stream line in hodograph plane is given by

$$\left(\frac{dv}{du}\right)_\psi = \frac{v\chi_{uu} - u\chi_{uv}}{-v\chi_{uv} + u\chi_{vv}} \tag{31}$$

At the limiting hodograph, Eq. (23) holds, therefore, Eq. (31) together with Eq. (28) gives

$$\left(\frac{dv}{du}\right)_{\psi,l} = -\left(\frac{\chi_{uv}}{\chi_{vv}}\right)_l = \frac{1-\frac{u^2}{a^2}}{\frac{uv}{a^2} \mp \sqrt{\frac{q^2}{a^2}-1}} \tag{32}$$

where the sign before the radical can be either negative or positive corresponding to the sign in Eq. (28).

On the other hand, the slope of the characteristics in hodograph plane is determined by Eq. (27). By solving for $\left(\frac{dv}{du}\right)_f$ and simplifying the result with aid of Eq. (15),

$$\left(\frac{dv}{du}\right)_f = \frac{\frac{uv}{a^2} - \frac{v}{\chi_v}\chi_{uv} \pm \sqrt{\frac{q^2}{a^2}-1}}{\left(1-\frac{v^2}{a^2}\right) + \frac{v}{\chi_v}\chi_{vv}} \tag{33}$$

The sign before the radical is either positive or negative corresponding to the two families of characteristics. By using the positive sign in conjunction with positive sign in Eq. (28), and similarly for the negative sign,

$$\left(\frac{dv}{du}\right)_{f,l} = \frac{1-\frac{u^2}{a^2}}{\frac{uv}{a^2} \mp \sqrt{\frac{q^2}{a^2}-1}} \tag{34}$$

Eqs. (32) and (34) show that the stream lines and one family of characteristics are tangent to each other at the limiting hodograph. This result is the same as that obtained for two-dimensional flow. (Ref. 7, 8, 9). These equations when compared with Eq. (12) for the slope of Mach waves in physical plane yields the interesting result that the stream lines and one family of characteristics at limiting hodograph are perpendicular to the corresponding Mach waves at the limiting line.

Since

$$\left(\frac{dv}{du}\right)_\psi = -\frac{\psi_u}{\psi_v} \tag{35}$$

Eq. (32) gives the following equation which holds at the limiting hodograph

$$\left(1-\frac{v^2}{a^2}\right)(\psi_u)_\ell^2 + 2\frac{uv}{a^2}(\psi_u)_\ell(\psi_v)_\ell + \left(1-\frac{u^2}{a^2}\right)(\psi_v)_\ell^2 = 0 \tag{36}$$

This equation can be reduced to more familiar form by introducing the polar coordinates in u, v plane:

$$u = q\cos\theta, \quad v = q\sin\theta$$

where θ is the angle between the velocity vector and u-axis. Then Eq. (36) takes the form

$$(\psi_q)_\ell^2 + \left(\frac{1}{q^2}-\frac{1}{a^2}\right)(\psi_\theta)_\ell^2 = 0 \tag{37}$$

This can be regarded as the equivalent to Eq. (23) for defining the limit hodograph. Similar relation exists for two dimensional flow. (Ref. 7, 8, 9)

Along a stream line, the ratio between $(dv)_\psi$ and $(du)_\psi$ is given by Eq. (31). By substituting this ratio into Eq. (19), the differential $(dx)_\psi$ and $(dy)_\psi$ along a stream line is given as

$$(dx)_\psi = \frac{u\left[\chi_{uu}\chi_{vv} - \chi_{uv}^2\right]}{-v\chi_{uv} + u\chi_{vv}}(du)_\psi \tag{38}$$

$$(dy)_\psi = \frac{v\left[\chi_{uu}\chi_{vv} - \chi_{uv}^2\right]}{-v\chi_{uv} + u\chi_{vv}}(du)_\psi$$

At the limiting line, Eq. (23) is satisfied. Then Eq. (38) shows that at the limiting line, the stream line has a singularity. Or, more plainly, $(dx)_\psi$

and $(dy)_\psi$ at these points are infinitesimals of higher order than $(du)_\psi$ and $(dv)_\psi$. By writing s for the distance measured along a stream line, Eq. (38) gives immediately

$$(u_s)_\psi = \frac{-v\chi_{uv} + u\chi_{vv}}{q[\chi_{uu}\chi_{vv} - \chi_{uv}^2]} \tag{39}$$

Similarly,

$$(v_s)_\psi = \frac{v\chi_{uu} - u\chi_{uv}}{q[\chi_{uu}\chi_{vv} - \chi_{uv}^2]} \tag{40}$$

Therefore, at the limiting line, the accelerations along a stream line is infinitely large. Furthermore, since the pressure gradient $(p_s)_\psi$ along a stream line is

$$(p_s)_\psi = -\rho q q_s = -\rho[u(u_s)_\psi + v(v_s)_\psi] \tag{41}$$

the pressure gradient at the limiting line is also infinitely large.

Such infinite acceleration and pressure gradient lead one to suspect that the fluid is thrown back at the limiting line. In other words, the stream lines are doubled back at this line of singularity. To investigate whether this is true, the character of the relation $\chi_{uu}\chi_{vv} - \chi_{uv}^2 = 0$ along a stream line has to be determined. If the derivative of this expression along a stream line is not zero, then $\chi_{uu}\chi_{vv} - \chi_{uv}^2$ has only a simple zero at the intersection of the limiting line and the stream line. Consequently, the differentials $(dx)_\psi$ and $(dy)_\psi$ will change sign by passing through the limiting hodograph in u-v plane along a stream line. Hence, the stream lines will double back and form a cusp at the limiting line. The derivative of $\chi_{uu}\chi_{vv} - \chi_{uv}^2$ along the stream line can be calculated with the aid of Eq. (30)

$$\left[\frac{d}{du}\left(\chi_{uu}\chi_{vv}-\chi_{uv}^{2}\right)\right]_{l}=\chi_{uuu}\chi_{vv}-2\chi_{uv}\chi_{uuv}+\chi_{uu}\chi_{uvv}$$
$$+\frac{v\chi_{uu}-u\chi_{uv}}{-v\chi_{uv}+u\chi_{vv}}\left\{\chi_{uuv}\chi_{vv}-2\chi_{uv}\chi_{uvv}+\chi_{uu}\chi_{vvv}\right\} \tag{42}$$

The expression on the right of Eq. (42) cannot be reduced to zero by the available relations which consists of Eq. (23), Eq. (15) and differentiated forms of Eq. (15). Therefore, the expression concerned generally only has a simple zero at the limiting hodograph and the stream lines are doubled back at the limiting line. It will be shown later that there is no solution possible beyond the limiting line. Hence, the name limiting line.

Envelope of Characteristics in Hodograph Plane and Lines of Constant Velocity in Physical Plane

Since the limiting line is the envelope of the Mach waves in physical plane, it is interesting to see whether there is also envelope for the characteristics in hodograph plane. The characteristics in u-v plane are determined by Eq. (26). The envelope to them can be found by eliminating $\left(\frac{dv}{du}\right)_f$ between Eq. (26) and the following equation

$$\left\{\left(1-\frac{v^2}{a^2}\right)+\frac{v}{\chi_v}\chi_{vv}\right\}\left(\frac{dv}{du}\right)_f-\left\{\frac{uv}{a^2}-\frac{v}{\chi_v}\chi_{uv}\right\}=0 \tag{43}$$

which is obtained by equating to zero the partial derivative of Eq. (26) with respect to $\left(\frac{dv}{du}\right)_f$. The result can be simplified by Eq. (15), and then it is simply

$$1-\frac{u^2+v^2}{a^2}+\frac{u^2v^2}{a^4}=\frac{u^2v^2}{a^4} \tag{44}$$

This is satisfied by either

$$a=0 \tag{45}$$

or

$$u^2+v^2=a^2 \tag{46}$$

The first condition, Eq. (45), when substituted into Eq. (26) gives

$$\left(\frac{dv}{du}\right)_{f,\ a=0} = -\frac{u}{v} \tag{47}$$

which shows that the circle of maximum velocity corresponding to a = 0, is the envelope to the characteristics in hodograph plane. The second condition, Eq. (46) is the spurious solution, since generally the characteristic at q = a does not tangent to the circle q = a. Hence a = 0 is the only envelope.

The lines of constant velocity in hodograph plane are simply circles. Therefore

$$\left(\frac{dv}{du}\right)_{q} = -\frac{u}{v} \tag{48}$$

By means of this relation and Eq. (19), the slope of the lines of constant velocity is given as

$$\left(\frac{dy}{dx}\right)_{q} = \frac{v\chi_{uv} - u\chi_{vv}}{v\chi_{uu} - u\chi_{uv}} \tag{49}$$

This equation together with Eq. (30) gives the following interesting relation

$$\left(\frac{dy}{dx}\right)_{q} = -\frac{1}{\left(\frac{dv}{du}\right)_{\psi}} \tag{50}$$

In other words, a line of constant velocity in physical plane is perpendicular to the stream line in hodograph plane at corresponding point.

The Lost Solution

Throughout the previous calculation, the possibility of using the Legendre transformation is assumed. This requires that for each pair of values of u, v there is one and only one pair of values of x, y. However, it is not always true, it is possible to have a number of points in the physical plane having the same value of u and v. If this is the case, then evidently it is impossible to solve for x and y from the pair of functions u = u (x,y), v = v (x,y). Mathematically, the situation is expressed by saying that the Jacobin $\partial(u,v)/\partial(x,y)$ vanishes in the physical plane. Or

$$u_x v_y - u_y v_x = 0 \tag{51}$$

However, this is also the condition for a functional relation between u and v, e.g., v can be expressed as a function of u. In other words, u and v are not independent. Hence if a solution is "lost" or not included in the family of solutions allowing Legendre transformation, then for that solution,

$$v = v(u) \tag{52}$$

It is seen that Eq. (51) is then identically satisfied.

By eliminating ρ from the continuity equation, we obtain

$$u_x\left(1-\frac{u^2}{a^2}\right) - \frac{uv}{a^2}\left(u_y + v_x\right) + \left(1-\frac{v^2}{a^2}\right)v_y + \frac{v}{y} = 0 \tag{53}$$

This equation can be rewritten in the following form by using Eq. (52)

$$\left\{\left(1-\frac{u^2}{a^2}\right) - \frac{uv}{a^2}\frac{dv}{du}\right\}u_x + \left\{\left(1-\frac{v^2}{a^2}\right)\frac{dv}{du} - \frac{uv}{a^2}\right\}u_y + \frac{v}{y} = 0 \tag{54}$$

The vorticity equation, Eq. (4) can be expressed as

$$\frac{dv}{du}u_x - u_y = 0 \tag{55}$$

From Eqs. (54) and (55), one can solve for u_x and u_y . The result is

$$\left[\left(1-\frac{u^2}{a^2}\right) - 2\frac{uv}{a^2}\frac{dv}{du} + \left(1-\frac{v^2}{a^2}\right)\left(\frac{dv}{du}\right)^2\right]u_x = -\frac{v}{y} \tag{56}$$

$$\left[\left(1-\frac{u^2}{a^2}\right) - 2\frac{uv}{a^2}\frac{dv}{du} + \left(1-\frac{v^2}{a^2}\right)\left(\frac{dv}{du}\right)^2\right]u_y = -\frac{v}{y}\frac{dv}{du} \tag{56b}$$

By differentiating the first of Eq. (56) with respect to y, the second with respect to x, the following relation is obtained by subtraction:

$$\frac{d^2v}{du^2}u_x + \frac{1}{y} = 0 \tag{57}$$

Therefore

$$\frac{dv}{du} = \frac{f(y) - x}{y} \tag{58}$$

or

$$y = \frac{f(y) - x}{\frac{dv}{du}}$$

where $f(y)$ is an undetermined function of y. However, Eq. (55) shows that for lines of constant values of u where $du = u_x\,(dx)_u + u_y\,(dy)_u = 0$

$$\left(\frac{dy}{dx}\right)_u = -\frac{1}{\left(\frac{dv}{du}\right)_u} = \text{constant} \tag{59}$$

Hence, lines of constant values of u and v are straight lines. This restriction reduces the function $f(y)$ in Eq. (58) to a numerical constant. Put $f(y) = K$, Eq. (58) is then

$$y = \frac{K - x}{\frac{dv}{du}} \tag{60}$$

Therefore lines of constant value of u and v are radial lines passing through the point x = K. Thus the lost solution is nothing but the well-known solution for the flow over a conical surface.

From Eq. (59), it is seen that lines of constant velocity are perpendicular to the tangent of u-v curve at the corresponding points. By substituting the value of $\frac{1}{y}$ from Eq. (57) into Eq. (56a), a relation between u and v is obtained:

$$v\frac{d^2v}{du^2} - \left(1 - \frac{v^2}{a^2}\right)\left(\frac{dv}{du}\right)^2 + 2\frac{uv}{a^2}\frac{dv}{du} - \left(1 - \frac{u^2}{a^2}\right) = 0 \tag{61}$$

This is the differential equation for the determining the hodograph representing the flow over a cone. Fig. 3 shows the hodograph for a cone of 30° semi-vertex angle and with a velocity at surface of cone equal to 0.35 c. c is the maximum velocity, i.e., the value of q corresponding to a = 0. Fig. 3 is drawn from data given by G. I. Taylor and J. W. Maccoll (Ref. 12).

It may well be mentioned here that the lost solution for the axially symmetric flow is not limited to supersonic velocity only as is the case for two-dimensional flow. In fact, Taylor and Maccoll show that for small forward velocity of the cone, supersonic velocities occur only just after the head shock wave. The velocity decreases as the surface of the cone is approached. Finally, it becomes subsonic for points near the surface of the cone. Fig. 4 shows a few examples taken from their calculations (Ref. 12). The dotted curves in the figure are the Mach waves. The dotted straight lines are the boundaries between the supersonic and the subsonic regions. Furthermore, spark photographs of conical shell in actual flight taken by Macc 11 (Ref. 13) do not indicate the presence of shock waves in regions of flow where such transition from supersonic to subsonic velocities is expected. Therefore, at least for this particular type of flow, a smooth transition through sonic velocity actually takes place.

Continuation of Solution Beyond the Limiting Line

Since it is shown in a previous paragraph that the stream lines are generally turned back at the limiting line, the question arises: Is it possible to continue the solution beyond the limiting line? Of course, there are two ways of continuing the solution: The new solution is joined either smoothly to the given solution at the limiting line or with a discontinuity. As shown before, the limiting line is the envelope of one family of the Mach waves, then at every point of this line its direction differs from that of stream line by an angle equal to the Mach angle. But the Mach angle is not zero except at points where the velocity of fluid has reached the maximum velocity and the ratio $\frac{a}{q}=0$. Therefore, the limiting line generally does not coincide with the stream line, and the discontinuity at the junction of the solution at the limiting line cannot be that of a vortex sheet. The only other type of discontinuity is the shock wave. However, the angle between the limiting line and flow direction is equal to Mach angle. Then according to the result of the theory of shock waves, the discontinuity across such a line vanishes. In other words, there cannot be a discontinuit at the limiting line. Therefore, it is impossible to join a new solution at the limiting line with a discontinuity.

As to the second possibility of joining a new solution smoothly at the limiting line, it is seen that the flow beyond the limiting line must be irrotational and isentropic since the limiting line cannot be a shock wave. There are only two types of isentropic irrotational flow, namely, one that allows the Legendre transformation and one that does not, the lost solution. Investigate the second alternative first. If the solution beyond the limiting line belongs to the so-called "lost solution", then since the junction at the limiting line must be smooth, the values of u, and v at the limiting line must also satisfy the Eq. (61). But, the slope $\left(\frac{dv}{du}\right)_l$ at limiting line is given by Eq. (24). The second derivative $\left(\frac{d^2v}{du^2}\right)_l$ will then involve the fourth order derivatives of χ. Besides these expressions, the available relations are Eqs. (23), (15), (25a) (25b) and three more equations obtained by differentiating Eqs. (25) with respect to u and v. However, it is still impossible for $\left(\frac{dv}{du}\right)_l$ to satisfy an equation like Eq. (61) where no derivative of χ appears. Hence, the limiting hodograph does not satisfy the equation for lost solution. In other words, the "lost solution" cannot be used to continue the flow beyond the limiting line.

The only remaining possibility is to continue the flow smoothly by another solution obtainable by Legendre transformation. Smooth continuation means that the values of u, v and ρ must be the same at the junction, the limiting line. Since shock waves do not appear, isentropic relations still hold. The density ρ is determined by velocity only. The value of u, and v are determined by the coordinates in hodograph plane. The position of the limiting line in physical is determined by χ_u, χ_v . Therefore, the problem can be stated as follows: At a certain given curve $u(\lambda)$, $v(\lambda)$ in the hodograph plane, the limiting hodograph, the values of χ_u, χ_v are given. λ is the parameter along the given curve. It is required to determine a new solution of the differential equation Eq. (15) with these initial values. First of all, it is seen that with the given data, the left hand sides of the following equations are given:

$$\frac{d}{d\lambda}(\chi_u) = \chi_{uu}\frac{du}{d\lambda} + \chi_{uv}\frac{dv}{d\lambda} \tag{62a}$$

$$\frac{d}{d\lambda}(\chi_v) = \chi_{uv}\frac{du}{d\lambda} + \chi_{vv}\frac{dv}{d\lambda} \tag{62b}$$

Therefore

$$\chi_{uv} = \left[-\frac{dv}{d\lambda}\chi_{vv} + \frac{d}{d\lambda}(\chi_v)\right]\Big/\frac{du}{d\lambda} \tag{63a}$$

$$\chi_{uu} = \left[\left(\frac{dv}{d\lambda}\right)^2\chi_{vv} - \frac{dv}{d\lambda}\frac{d}{d\lambda}(\chi_v) + \frac{du}{d\lambda}\frac{d}{d\lambda}(\chi_u)\right]\Big/\left(\frac{du}{d\lambda}\right)^2 \tag{63b}$$

By substituting those values into Eq. (15), the second degree terms reduce to

$$\chi_{uu}\chi_{vv} - \chi_{uv}^2 = \left[\frac{dv}{d\lambda}\frac{d}{d\lambda}(\chi_v) + \frac{du}{d\lambda}\frac{d}{d\lambda}(\chi_u)\right]\chi_{vv}\Big/\left(\frac{du}{d\lambda}\right)^2 + \left[\frac{d}{d\lambda}(\chi_v)\right]^2\Big/\left(\frac{du}{d\lambda}\right)^2 \tag{64}$$

which is linear in χ_{vv}. Therefore χ_{vv} can be uniquely determined by Eq. (15). In other words, with the given data, the second order derivatives of χ at the given curve $u(\lambda), v(\lambda)$ can be determined uniquely, in spite of the fact that the differential equation (15) is of second degree. K. Friedrichs and H. Lewy (Ref. 13) have shown that under these circumstances, the function χ within a region R (Fig. 5) bounded by two characteristics and the given curve is uniquely determined except on additional constant. Consequently there can be only one solution corresponding to the given data at the limiting hodograph. However, this solution is the very one which gives the reverse flow at limiting line. Therefore, it is impossible to continue the solution beyond the limiting line even by Legendre transformation.

Since all three alternatives fail to offer a way of continuing the solution, the limiting line is truly an impossible boundary to cross. In other words, the region beyond the limiting line is a "forbidden region". This physical absurdity can only be resolved by the breakdown of isentropic irrotational flow some distance ahead of the limiting line.

General Three Dimensional Flow

The methods used in previous sections for investigating the axially symmetric flow can be easily extended to the general three dimensional case. In the present section, this investigation will be sketched briefly and the results indicated.

Let the three components of velocity along the three coordinate axes x, y and z be denoted by u, v, and w respectively. Then by introducing a velocity potential φ defined by

$$u = \varphi_x , \quad v = \varphi_y , \quad w = \varphi_z \tag{65}$$

the differential equation for φ of an isentropic irrotational flow can be written as (Ref. 7)

$$a^2(\varphi_{xx} + \varphi_{yy} + \varphi_{zz}) = u^2\varphi_{xx} + v^2\varphi_{yy} + w^2\varphi_{zz} + 2vw\varphi_{yz} + 2wu\varphi_{zx} + 2uv\varphi_{xy} \tag{66}$$

If for every triad of u, v, w, there is only one triad of x, y, z, then the Legendre transformation can be used. Thus

$$\chi = ux + vy + wz - \varphi \tag{67}$$

and

$$\chi_u = x, \quad \chi_v = y, \quad \chi_w = z \tag{68}$$

The differential equation for φ, Eq. (66), is then transformed into

$$\begin{aligned} a^2[BC - F^2 + CA - G^2 + AB - H^2] &= u^2(BC - F^2) + v^2(CA - G^2) \\ &+ w^2(AB - H^2) + 2vw(GH - AF) + 2wu(HF - BG) + 2uv(FG - CH) \end{aligned} \tag{69}$$

where the following notations are used

$$A = \chi_{uu},\quad B = \chi_{vv},\quad C = \chi_{ww},\quad F = \chi_{vw},\quad G = \chi_{wu},\quad H = \chi_{uv} \tag{70}$$

By analogy with the axially symmetric case, the limiting hodograph surface is defined as the surface in the u, v, w space, or hodograph space, where the following relation holds:

$$\Delta = \begin{vmatrix} A & H & G \\ H & B & F \\ G & F & C \end{vmatrix} = 0 \tag{71}$$

The properties of this limiting hodograph and the corresponding limiting surface can be found by considering the behavior of stream lines and characteristics at such surfaces.

From Eq. (68), the differentials of x, y and z can be written as

$$dx = A\,du + H\,dv + G\,dw \tag{72a}$$

$$dy = H\,du + B\,dv + F\,dw \tag{72b}$$

$$dz = G\,du + F\,dv + C\,dw \tag{72c}$$

Along a stream line, the differentials dx, dy and dz must be proportional to u, v and w respectively. Thus the equation of a stream line in physical space is

$$\frac{(dx)_\psi}{u} = \frac{(dy)_\psi}{v} = \frac{(dz)_\psi}{w} \tag{73}$$

where the subscript ψ indicates values taken along the stream line. The equation of a stream line in hodograph space is obtained by eliminating dx, dy and dz from Eq. (73) by Eq. (72). The result is

$$\frac{(du)_\psi}{\bar{a}u + \bar{h}v + \bar{g}w} = \frac{(dv)_\psi}{\bar{h}u + \bar{b}v + \bar{f}w} = \frac{(dw)_\psi}{\bar{g}u + \bar{f}v + \bar{c}w} \tag{74}$$

where $\bar{a}$ is the co-factor of A in the determinante Δ of Eq. (71), $\bar{b}$ the co-factor [redacted] etc. Eq. (74) can be used in turn to eliminate two of the three differentials du, dv and dw in the right of Eq. (72). The result is

$$(dx)_\psi = \frac{u\Delta\, du}{\bar{a}u + \bar{h}v + \bar{g}w} \tag{75a}$$

$$(dy)_\psi = \frac{v\Delta\, dv}{\bar{h}u + \bar{b}v + \bar{f}w} \tag{75b}$$

$$(dz)_\psi = \frac{w\Delta\, dw}{\bar{g}u + \bar{f}v + \bar{c}w} \tag{75c}$$

At the limiting surface, $\Delta = 0$ as defined by Eq. (71), therefore the stream lines have a singularity there. Similar to the axially symmetric flow, the stream lines generally are turned back and form a cusp at this surface. The acceleration and the pressure gradient are, of course, infinitely large at such places.

The characteristic surface $g(x, y, z) = 0$ in physical space is determined by the equation

$$a^2[g_x^2 + g_y^2 + g_z^2] = u^2 g_x^2 + v^2 g_y^2 + w^2 g_z^2 + 2vw\, g_y g_z + 2wu\, g_z g_x + 2uv\, g_x g_y \tag{76}$$

Since this equation is a second degree equation, there are two families of surfaces passing through each point. These surfaces are the wave fronts of infinitesimal distrubances in the flow and can be called the Mach surfaces. The characteristic surface $f(u, v, w) = 0$ in the hodograph space is determined by the equation

$$\begin{aligned}
& a^2[(B+C)f_u^2 + (C+A)f_v^2 + (A+B)f_w^2 - 2F f_v f_w - 2G f_w f_u - 2H f_u f_v] \qquad (77)\\
&= u^2[C f_v^2 + B f_w^2 - 2F f_v f_w] + v^2[C f_u^2 + A f_w^2 - 2G f_w f_u]\\
&+ w^2[B f_u^2 + A f_v^2 - 2H f_u f_v] + 2vw[H f_w f_u + G f_u f_v - F f_u^2 - A f_v f_w]\\
&+ 2wu[F f_u f_v + H f_v f_w - G f_v^2 - B f_w f_u]\\
&+ 2uv[G f_v f_w + F f_w f_u - H f_w^2 - C f_u f_v]
\end{aligned}$$

By transforming Eq. (76) for Mach surfaces to hodograph space, it can be shown that the transformed equation is satisfied by either the characteristics in hodograph space determined by Eq. (77) or the limiting hodograph determined by Eq. (71). Therefore, here again the limiting surface is the envelope of a family of Mach surfaces.

By using Eqs. (74) and (77), it is possible to show that the stream lines in hodograph space are tangent to the characteristic surfaces at the limiting hodograph. Furthermore, by using Eqs. (69), (71) and (74), the inclination of the stream lines at the limiting hodograph can be calculated. In fact, if $(ds)^2 = (du)^2 + (dv)^2 + (dw)^2$, $q^2 = u^2 + v^2 + w^2$ the following relation is obtained

$$\left(\frac{ds}{dq}\right)_{\psi, l} = \frac{q}{a} \quad \text{or} \quad -\frac{q}{a} \tag{78}$$

This relation is really equivalent to Eq. (32). In other words, at the limiting hodograph, the inclination of the stream lines and characteristics from the q = constant surface is equal to the Mach angle (Fig. 6). It thus seems the break down of general steady isentropic irrotational flow of non-viscous fluid is connected with the appearance of the envelope of Mach waves in physical space and the tangency of stream lines and characteristics in hodograph space.

REFERENCES

1. Theodorsen, T. "The Reaction on a Body in a Compressible Fluid" J. Aero. Sciences, Vol. 4, pp. 239-240, (1937).

2. Taylor, G. I., and Sharman, C. F., "A Mechanical Method for Solving Problems of Flow in Compressible Fluids" Proc. of Royal Society (A). Vol. 121, pp. 194-217, (1928).

3. Taylor, G. I., "Recent Works on the Flow of Compressible Fluids" Journal London Mach. Society, Vol. 5, pp. 224-240. (1930).

4. Binnie, A. M. and Hooker, S. G., "The Radial and Spiral Flow of a Compressible Fluid." Philosophical Magazine, (7), Vol. 23, pp. 597-606, (1937).

5. Tollmien, W., "Zum Übergang von Unterschall - in Überschall Strömungen, Z.a.M.M., Vol. 17, pp. 117-136, (1937).

6. Ringleb, F., "Exakte Lösungen der Differentialgleichungen einer adiabatischen Gasströmung, Ibed, Vol. 20, pp. 185-198, (1940)

7. von Kármán, Th., "Compressibility Effects in Aerodynamics" J. Aero. Sciences, Vol. 8, pp. 337-356, in particular pp. 351-356, (1941).

8. Ringleb, F., "Über die Differentialgleichungen einer adiabatichen Gasströmung und den Strömungsstoss. Deutsche Mathematik, Vol. 5, pp. 377-384 (1940).

9. Tollmien, W., "Grenzlinien adiabatischer Potentialstromungen" Z.a.M.M., Vol. 21, pp. 140-152 (1941).

10. Frankle, F., Bulletin de l'academie des sciences, U.R.S.S. (7), 1934.

The convergence proof is given by Frankle, F. and Aleksejeva, R., "Zwei Randwertaufgaben aus der Theorie der hyperbolischen partiellen Differentialgleichungen zweiter Ordnung mit Anwendungen auf Gasströmungen mit Überschallgeschwindigkeit." Matematiceski Sbornik, Vol. 41, pp. 483-502, 1935, (Russian, German summary).

11. Ferrari, C., "Campo aerodinamico a velocita iperacustica attorno a un solido di revoluzione a prora acuminata." L'Aerotecnica, Vol. 16, pp. 121-130 (1936).

Also "Determinazione della pressione sopra solidi di revoluzione a prora acuminata disposti in deriva in corrente di fluido compressibile a velocita ipersonora" Atti della R. Accademia delle Scienze di Torino, Vol. 72, pp. 140-163 (1937).

12. Taylor, G. I. and Maccoll, J. W. "The Air Pressure on a cone moving at High Speeds." Proc. Royal Society (A), Vol. 139, pp. 278-298 (1933).

13. Maccoll, J. W. "The Conical Shock Wave Formed by a Cone Moving at a High Speed." Ibid. Vol. 159, pp. 459-472 (1936).

14. Friedrichs, K. and Lewy, H., "Das Anfangswertproblem einer beliebigen nichtlinearen hyperbolischen Differentialgleischung belieliger Ordnung in zwei Variablen." Math. Annalen, Vol. 99, pp. 200-221, (1928).

Figure Captions

Fig. 1 Limiting Line as the Envelope of Mach Waves

Fig. 2 Stream Line and Velocity Components in an axially symmetric Flow

Fig. 3 Hodograph of the Flow over a Cone of 30° Half Vertex Angle and a Surface Velocity q equal to 0.35C.

Fig. 4 Flow over Cones of Various Vertex Angle involving Subsonic Regions. θ_s = Half Vertex Angle, u_s = Velocity over the Surface of Cone

Fig. 5 Region R where the Solution is uniquely determined by given Data at the Limiting Hodograph.

Fig. 6 Geometrical Relations between Stream Line and Characteristic Surface at Limiting Surface in Hodograph Space.

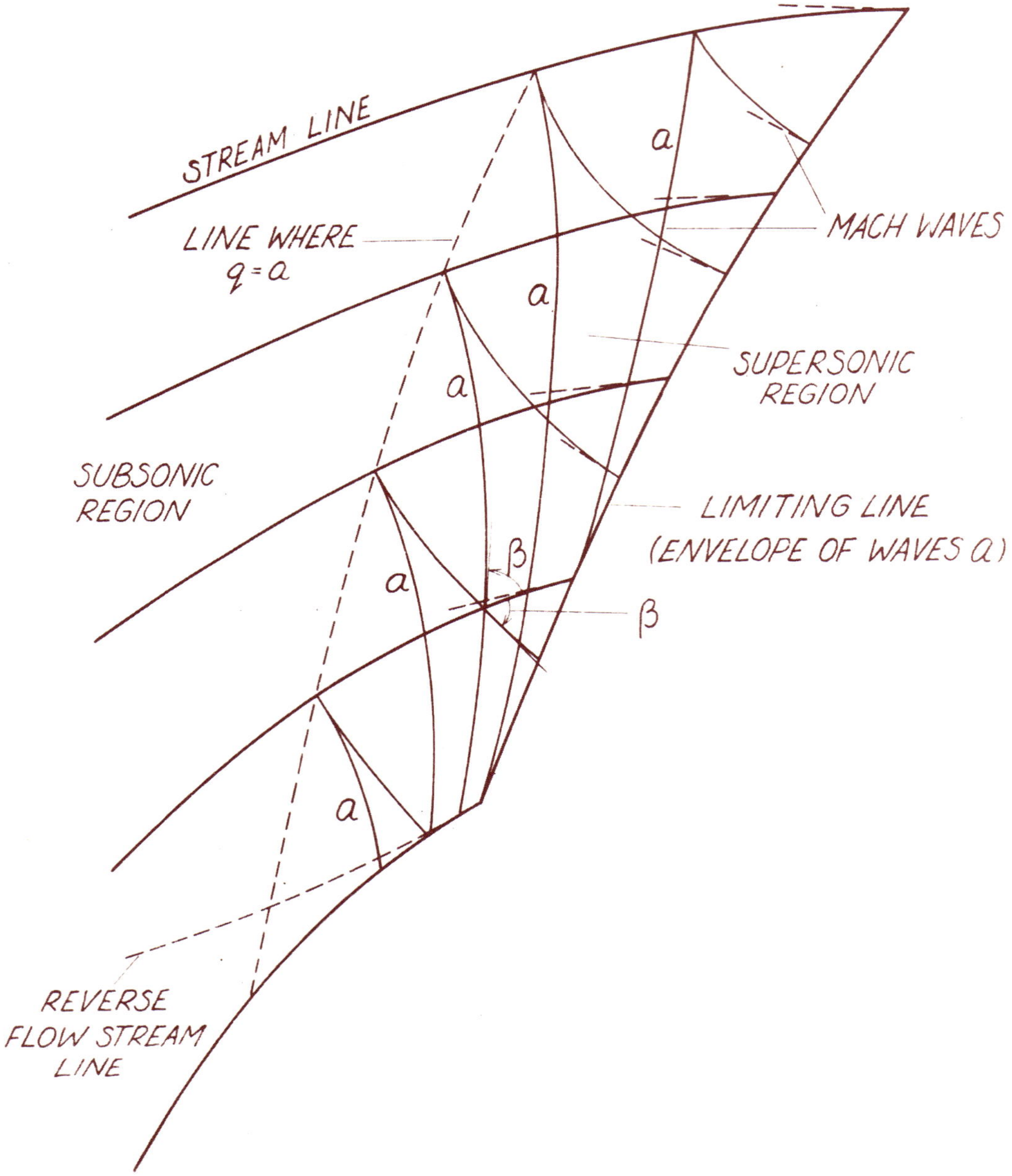

STREAM LINE
LINE WHERE
q=a
MACH WAVES
a
a
a
a
a
SUPERSONIC
REGION
SUBSONIC
REGION
LIMITING LINE
(ENVELOPE OF WAVES a)
β
β
REVERSE
FLOW STREAM
LINE

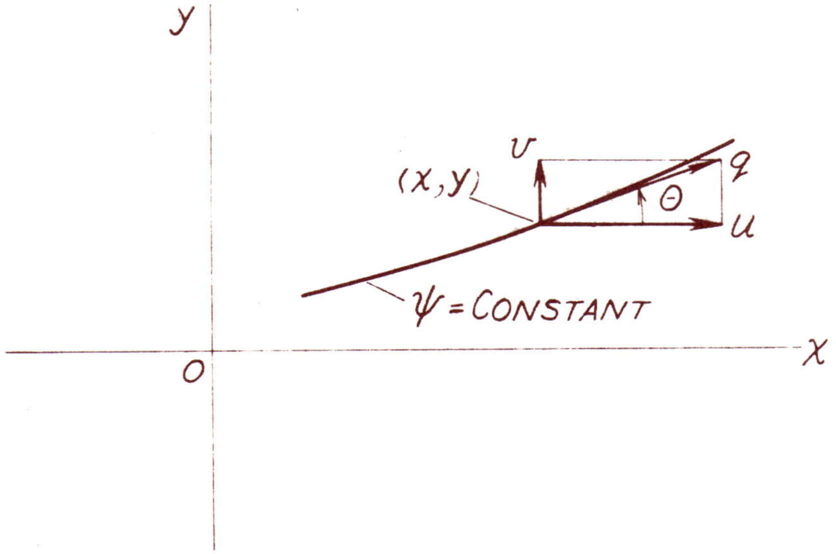
y
v
(x, y)
q
θ
u
ψ = CONSTANT
O
x

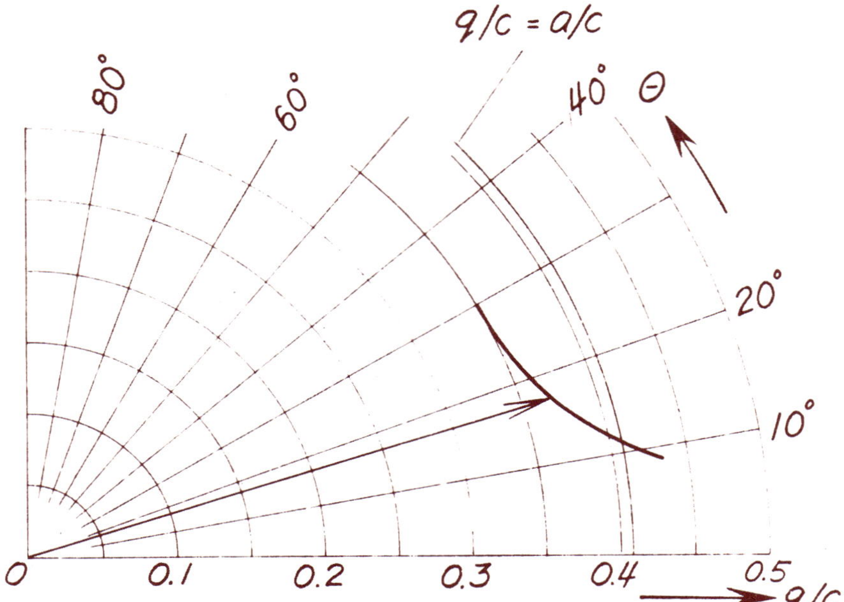

HODOGRAPH PLANE

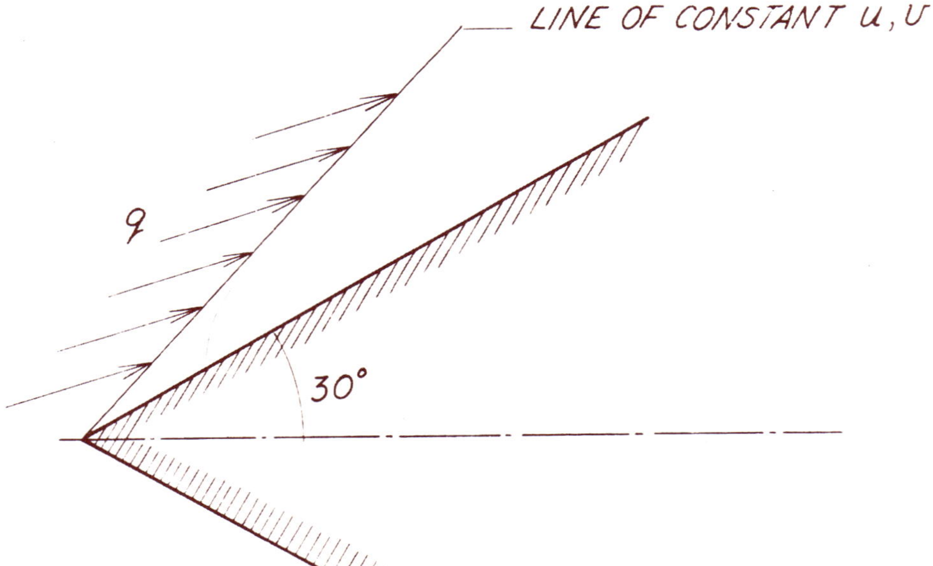

PHYSICAL PLANE

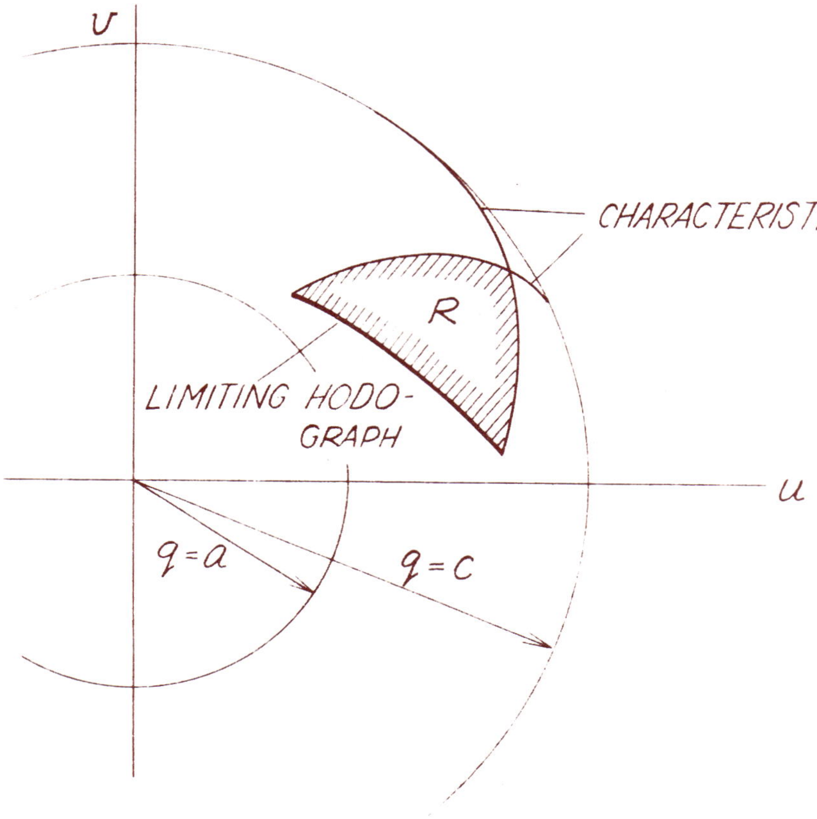
v
CHARACTERISTICS
R
LIMITING HODO-
GRAPH
u
q=a
q=c

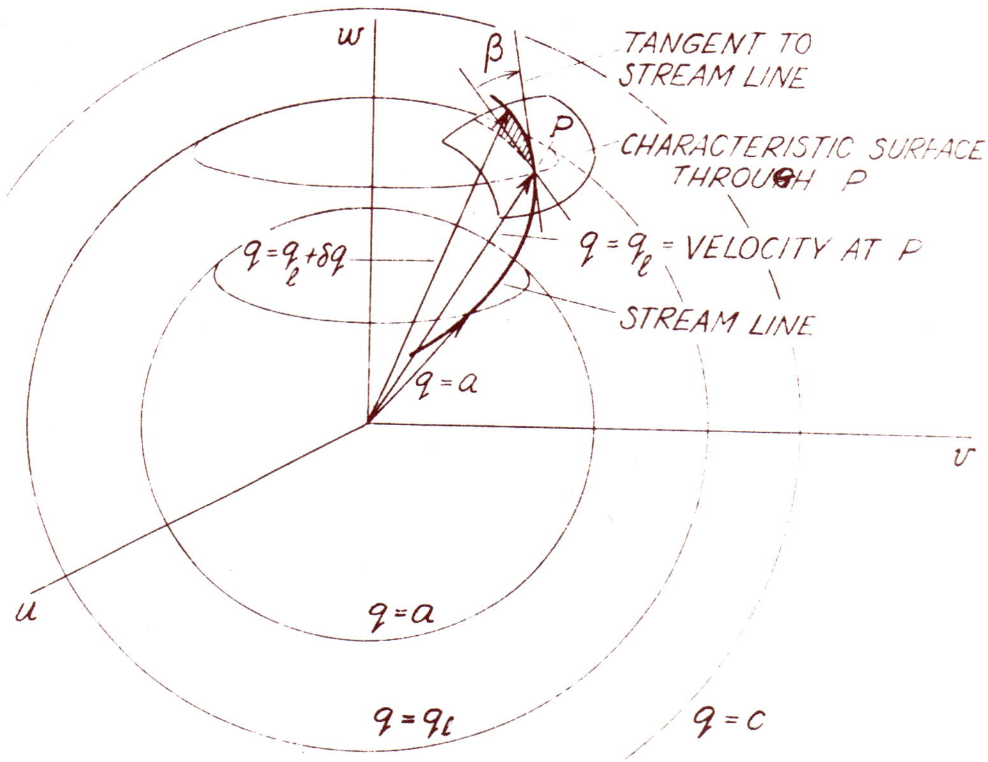
w
β
TANGENT TO
STREAM LINE
P
CHARACTERISTIC SURFACE
THROUGH P
q = q_l + δq
q = q_l = VELOCITY AT P
STREAM LINE
q = a
v
u
q = a
q = q_l
q = c

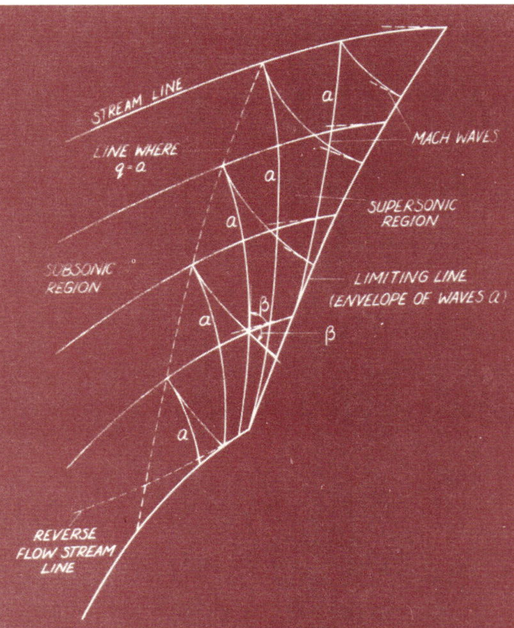

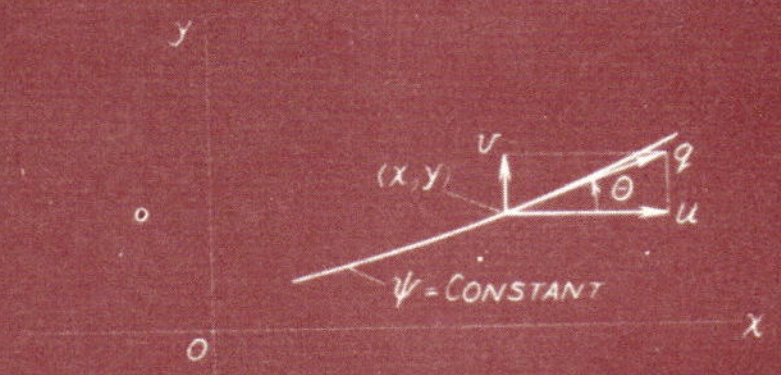

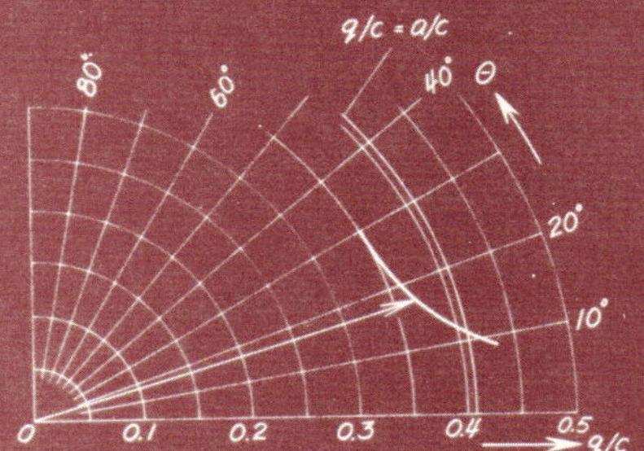

HODOGRAPH PLANE

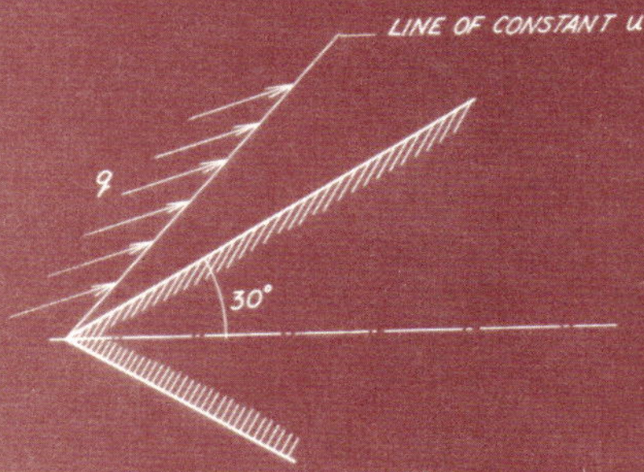

PHYSICAL PLANE

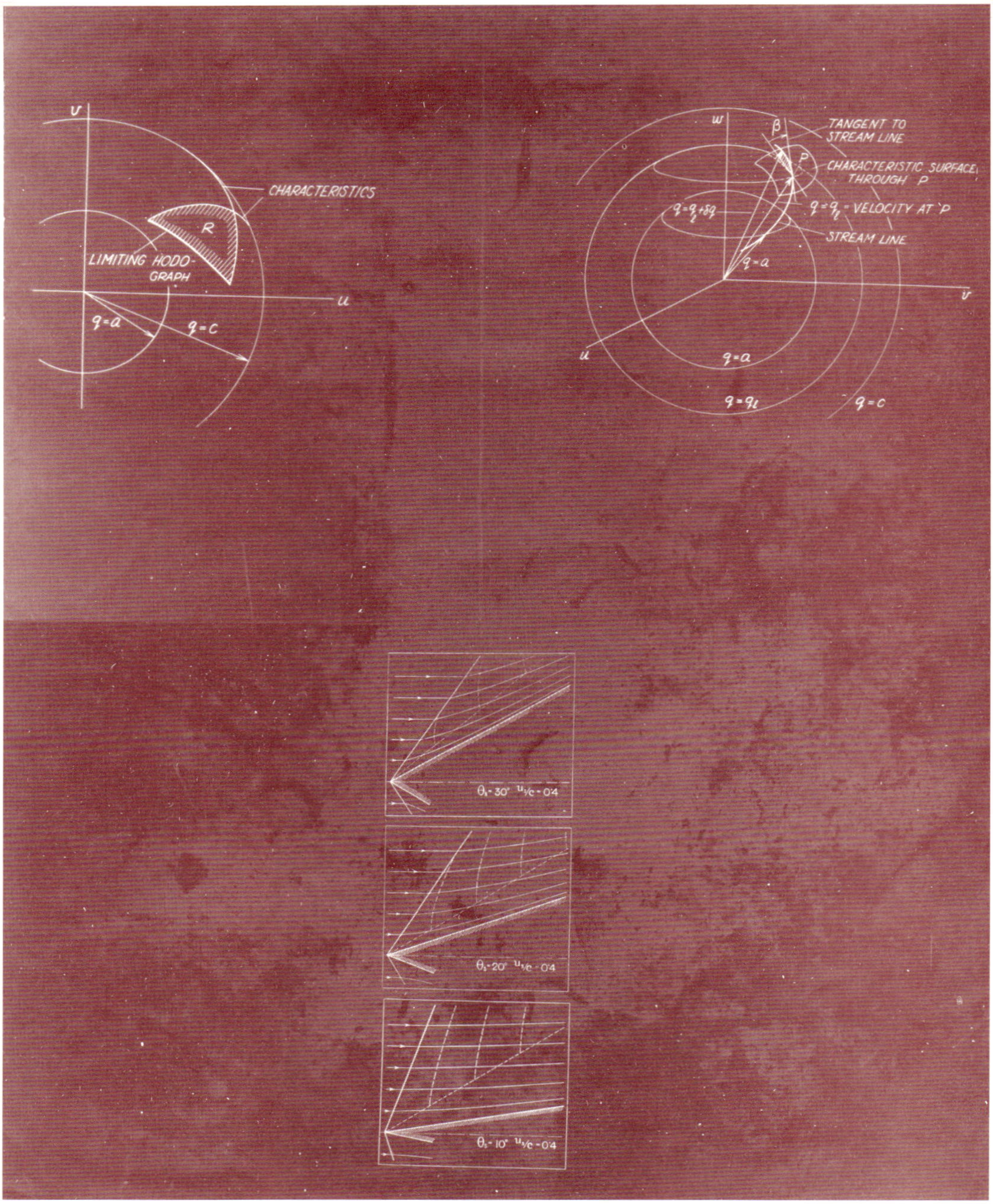
v
CHARACTERISTICS
R
LIMITING HODO-
GRAPH
u
q=a
q=c
w
TANGENT TO
STREAM LINE
β
P
CHARACTERISTIC SURFACE
THROUGH P
q=q1+δq
q=q1 = VELOCITY AT P
STREAM LINE
q=a
v
u
q=a
q=q1
q=c
θ1=30° u1/c=0·4
θ1=20° u1/c=0·4
θ1=10° u1/c=0·4

QUARTERLY OF APPLIED MATHEMATICS

BROWN UNIVERSITY · PROVIDENCE, R. I.

BOARD OF EDITORS:
Hugh L. Dryden
NATIONAL BUREAU OF STANDARDS
WASHINGTON, D. C.
Thornton C. Fry
BELL TELEPHONE LABORATORIES
463 WEST ST., NEW YORK, N. Y.
Theodore von Kármán
CALIFORNIA INSTITUTE OF TECHNOLOGY
PASADENA, CALIF.
John M. Lessells
MASSACHUSETTS INSTITUTE OF TECHNOLOGY
CAMBRIDGE, MASS.
Ivan S. Sokolnikoff
UNIVERSITY OF WISCONSIN
MADISON, WIS.
John L. Synge
UNIVERSITY OF TORONTO
TORONTO, CANADA
W. Prager, Managing Editor
BROWN UNIVERSITY
PROVIDENCE, R. I.

April 12, 1943

Dr. Hsue-Shen Tsien
California Institute of Technology
Pasadena, California

Dear Dr. Tsien:

I just received the program of the Meeting of the Mathematical Society at Stanford University on April 24, 1943. I see that you will address this meeting on The limiting line in mixed subsonic and supersonic flows. If you have not yet made other arrangements concerning the publication of this manuscript, I should like to suggest that you consider the Quarterly of Applied Mathematics as a possible means of publication.

Very sincerely yours,

W. Prager.

W. Prager

WP:G

UNIVERSITY OF MARYLAND
COLLEGE PARK

COLLEGE OF ARTS AND SCIENCES
DEPARTMENT OF MATHEMATICS

April 24, 1943

Dr. Hsne-Shen Tsien
California Institute of Technology
Pasaderra, California

Dear Dr. Tsien,

I have noted that you plan to speak upon "The limiting line in mixed subsonic and supersonic flows of compressible fluids" on April 24th.

This sounds very interesting to me in the light of some recent investigations I have undertaken, and I hope you will be kind enough to send me a reprint of your lecture when it becomes available.

Sincerely yours,

Monroe H. Martin

MHM:ES

Monroe H. Martin

May 7, 1943

Robert R. Dexter
Institute of the Aeronautical Sciences
1505 RCA Building West
30 Rockefeller Plaza
New York, New York

Dear Mr. Dexter:

Enclosed is my paper entitled "The Limiting Line in Mixed Subsonic and Supersonic Flow of Compressible Fluids." I am submitting it for consideration of publication in the Journal.

Sincerely yours

H. S. Tsien

HST:mo
Encl:

INSTITUTE
OF THE
AERONAUTICAL SCIENCES
OFFICIAL PUBLICATIONS
JOURNAL OF THE AERONAUTICAL SCIENCES
AERONAUTICAL ENGINEERING REVIEW
1505 RCA BUILDING WEST
30 ROCKEFELLER PLAZA
NEW YORK

President
Hugh L. Dryden
Past Presidents
J. C. Hunsaker
Charles L. Lawrance
D. W. Douglas
Glenn L. Martin
Clark B. Millikan
T. P. Wright
George W. Lewis
James H. Doolittle
Frank W. Caldwell
Hall L. Hibbard
Vice Presidents
J. L. Atwood
E. R. Breech
Sherman M. Fairchild
Earl D. Osborn
Executive Vice President
Lester D. Gardner
Secretary
Robert R. Dexter
Treasurer
Elmer A. Sperry, Jr.
Assistant Treasurer
George R. Forman
Controller
Joseph J. Maitan
COUNCIL
P. R. Bassett
W. A. M. Burden
Charles H. Colvin
R. T. Goodwin
L. S. Hobbs
C. S. Jones
Roger Wolfe Kahn
W. Wallace Kellett
James H. Kimball
John C. Leslie
R. D. MacCart
Arthur Nutt
ADVISORY BOARD
John D. Akerman
Lawrence D. Bell
Don R. Berlin
Lyman J. Briggs
William K. Ebel
Jack Frye
P. G. Johnson
Th. von Karman
J. H. Kindelberger
F. K. Kirsten
Paul Kollsman
Arnold M. Kuethe
William Littlewood
Thomas A. Morgan
Richard W. Palmer
W. A. Patterson
A. E. Raymond
F. W. Reichelderfer
J. T. Trippe
J. G. Vincent
Edward P. Warner

May 13, 1943

Dr. H. S. Tsien
Daniel Guggenheim Aeronautical Laboratory
California Institute of Technology
Pasadena, Calif.

Dear Dr. Tsien:

Thank you for submitting your paper "The Limiting Line in Mixed Subsonic and Supersonic Flow of Compressible Fluids" for possible publication in the "Journal". I am putting the paper through the usual procedure and will notify you of the comments of the Editorial Board as soon as possible.

Best regards

Sincerely yours,
Robert R. Dexter
Editor

GRF:MM

BULLETIN OF THE AMERICAN MATHEMATICAL SOCIETY

P. A. SMITH
MEMBER OF EDITORIAL COMMITTEE

COLUMBIA UNIVERSITY
NEW YORK, N. Y.

May 17, 1943

Dr. Hsue-Shen Tsien
California Institute of Technology
Pasadena, California

Dear Dr. Tsien:

I am writing with regard to your address, THE "LIMITING LINE" IN MIXED SUBSONIC AND SUPERSONIC FLOWS OF COMPRESSIBLE FLUIDS, delivered at the recent meeting of the American Mathematical Society at Stanford University. Would you care to offer a manuscript for possible publication in the Bulletin? I think we could allow about eight to ten printed pages.

As you probably know, papers in applied mathematics are subject to censorship and, therefore, there is some risk of having publication held up for an indefinite period of time. I hope, however, that in spite of this risk I shall receive a favorable reply from you.

Sincerely yours,

Paul A. Smith

Paul A. Smith

PAS:MF

INSTITUTE
OF THE
AERONAUTICAL SCIENCES
OFFICIAL PUBLICATIONS
JOURNAL OF THE AERONAUTICAL SCIENCES
AERONAUTICAL ENGINEERING REVIEW
1505 RCA BUILDING WEST
30 ROCKEFELLER PLAZA
NEW YORK

President
Hugh L. Dryden
Past Presidents
J. C. Hunsaker
Charles L. Lawrance
D. W. Douglas
Glenn L. Martin
Clark B. Millikan
T. P. Wright
George W. Lewis
James H. Doolittle
Frank W. Caldwell
Hall L. Hibbard
Vice Presidents
J. L. Atwood
E. R. Breech
Sherman M. Fairchild
Earl D. Osborn
Executive Vice President
Lester D. Gardner
Secretary
Robert R. Dexter
Treasurer
Elmer A. Sperry, Jr.
Assistant Treasurer
George R. Forman
Controller
Joseph J. Maitan
COUNCIL
P. R. Bassett
W. A. M. Burden
Charles H. Colvin
R. T. Goodwin
L. S. Hobbs
C. S. Jones
Roger Wolfe Kahn
W. Wallace Kellett
James H. Kimball
John C. Leslie
R. D. MacCart
Arthur Nutt
ADVISORY BOARD
John D. Akerman
Lawrence D. Bell
Don R. Berlin
Lyman J. Briggs
William K. Ebel
Jack Frye
P. G. Johnson
Th. von Karman
J. H. Kindelberger
F. K. Kirsten
Paul Kollsman
Arnold M. Kuethe
William Littlewood
Thomas A. Morgan
Richard W. Palmer
W. A. Patterson
A. E. Raymond
F. W. Reichelderfer
J. T. Trippe
J. G. Vincent
Edward P. Warner

July 22, 1943

Mr. Hsue-shen Tsien
Guggenheim Laboratory
California Institute of Technology
Pasadena, California

Dear Mr. Tsien:

Enclosed is your paper, "The 'Limiting Line' in Mixed Subsonic and Supersonic Flow of Compressible Fluids", which you submitted for publication in the "Journal". This paper was returned to us by the War Department with the following comment: "Objection is taken to unclassified publication of the article entitled, 'The Limiting Line in Mixed Subsonic and Supersonic Flow of Compressible Fluids'. This paper contains information that would be of aid to the enemy."

Your paper is an excellent one and was highly recommended for early publication by Dr. Dryden. I regret that we are unable to publish it at this time. Possibly, if you do not find means of publishing it as a classified report, you may be willing to submit it at some later date for publication in the "Journal".

Thank you for submitting the paper to us. I hope you will continue to send future papers for possible publication. I should appreciate your acknowledging the return of this material to you for our records.

Best regards.

Sincerely yours,

Robert R. Dexter

Secretary

RRD:BK
Enc.

September 14, 1944.

Mr. R. G. Robinson
National Advisory Committee for Aeronautics
1500 New Hampshire Ave., Dupont Circle
Washington 25, D. C.

Dear Mr. Robinson:

About a year and a half ago, I submitted to NACA a paper entitled "The Limiting Line in Mixed Subsonic and Supersonic Flow of Compressible Fluids" at the suggestion of Dr. von Karman for consideration of publication as a classified report. Six months ago, Dr. von Karman was in Washington and has talked about this matter with your staff, and found that the paper was mislaid. However, Dr. von Karman told me that it was decided to publish this paper at an early date.

Since another six months have passed, I wonder whether you will permit me to inquire again about the publication of the paper. As this paper discusses a subject of great current interest, its early publication should, perhaps, be considered.

I apologize for troubling you with such a trivial matter, and I remain

Very sincerely yours,

Hsue-shen Tsien.

JEROME C. HUNSAKER, SC. D., CHAIRMAN
LYMAN J. BRIGGS, PH. D., VICE CHAIRMAN
CHARLES G. ABBOT, SC. D.
GEN. HENRY H ARNOLD, U. S. A.
HON. WILLIAM A. M. BURDEN
VANNEVAR BUSH, SC. D.
WILLIAM F. DURAND, PH. D.
MAJ. GEN. OLIVER P. ECHOLS, U. S. A.
REAR ADMIRAL JOHN S. MCCAIN, U. S. N.
GEORGE J. MEAD, SC. D.
REAR ADMIRAL E. M. PACE U. S. N.
FRANCIS W. REICHELDERFER, SC. D.
EDWARD WARNER, SC. D.
ORVILLE WRIGHT, SC. D.
THEODORE P. WRIGHT, SC. D.

LANGLEY MEMORIAL AERONAUTICAL LABORATORY
LANGLEY FIELD, HAMPTON, VA.

AMES AERONAUTICAL LABORATORY
MOFFETT FIELD, CALIF.

AIRCRAFT ENGINE RESEARCH LABORATORY
CLEVELAND AIRPORT, CLEVELAND, OHIO

RESTRICTED

NATIONAL ADVISORY COMMITTEE
FOR AERONAUTICS

1500 NEW HAMPSHIRE AVE., DUPONT CIRCLE
WASHINGTON 25, D. C.

September 14, 1944

TELEPHONES: EXECUTIVE 3515 3516 3517

Dr. Hsue-Shen Tsien
California Institute of Technology
Pasadena, California

Dear Dr. Tsien:

With your letter of August 18, 1943 you forwarded to the Committee a copy of your paper "The 'Limiting Line' in Mixed Subsonic and Supersonic Flow of Compressible Fluids" for consideration of its release as a classified Committee report.

Study of this report by the Committee's technical staff has been completed and I am pleased to advise you that your paper will be published as a Restricted Technical Note of the Committee. If it appears advisable to refer the report to you in its final form prior to publication, and I suspect that this may be the case because of the amount of mathematical work included, the Committee would like to refer a proof copy to you.

Thank you for your interest in making the paper available to the Committee's staff and in offering it for publication.

Very truly yours,

NATIONAL ADVISORY COMMITTEE
FOR AERONAUTICS

G. W. Lewis
Director of
Aeronautical Research

RESTRICTED

RESTRICTED

September 18, 1944.

Dr. G. W. Lewis
Director of Aeronautical Research
National Advisory Committee for Aeronautics
1500 New Hampshire Ave., Dupont Circle
Washington 25, D. C.

Dear Dr. Lewis:

It is indeed a happy coincidence that only two days ago I wrote to Mr. R. G. Robinson inquiring about the publication of my paper "The 'Limiting Line' in Mixed Subsonic and Supersonic Flow of Compressible Fluids" and then I received your letter of September 14, 1944 confirming its publication as a Restricted Technical Note. Following your suggestion, I would like to see the report in its final form as the paper involved considerable mathematical work.

I thank you for informing me, and I remain

Very sincerely yours,

HST pl

Hsue-shen Tsien.

RESTRICTED

LANGLEY MEMORIAL AERONAUTICAL LABORATORY
LANGLEY FIELD, HAMPTON, VA.

AMES AERONAUTICAL LABORATORY
MOFFETT FIELD, CALIF.

AIRCRAFT ENGINE RESEARCH LABORATORY
CLEVELAND AIRPORT, CLEVELAND, OHIO

NATIONAL ADVISORY COMMITTEE FOR AERONAUTICS

1500 NEW HAMPSHIRE AVE., DUPONT CIRCLE
WASHINGTON 25, D. C.

TELEPHONES: EXECUTIVE 3515 3516 3517

September 23, 1944

Dr. Hsue-Shen Tsien
California Institute of Technology
Pasadena, California

Dear Dr. Tsien:

I now have more complete information on your paper "The 'Limiting Line' in Mixed Subsonic and Supersonic Flow of Compressible Fluids" and in transmitting it I wish to acknowledge your letter of September 18 and one of September 14 addressed to Mr. Robinson.

A mimeographed proof copy will be ready approximately three weeks from now and will be forwarded for your review before proceeding with the duplicating process. The Committee will be pleased to have your comments at that time and the paper will be published and released immediately thereafter.

Very truly yours,

NATIONAL ADVISORY COMMITTEE
FOR AERONAUTICS

R. G. Robinson

For G. W. Lewis
Director of
Aeronautical Research

RESTRICTED

RESTRICTED

REGISTERED MAIL

JEROME C. HUNSAKER, SC. D., CHAIRMAN
LYMAN J. BRIGGS, PH. D., VICE CHAIRMAN
CHARLES G. ABBOT, SC. D.
GEN. HENRY H. ARNOLD, U. S. A.
HON. WILLIAM A. M. BURDEN
VANNEVAR BUSH, SC. D.
WILLIAM F. DURAND, PH. D.
MAJ. GEN. OLIVER P. ECHOLS, U. S. A.
WILLIAM LITTLEWOOD, M. E.
VICE ADMIRAL JOHN S. MCCAIN, U. S. N.
REAR ADMIRAL E. M. PACE, U. S. N.
FRANCIS W. REICHELDERFER, SC. D.
EDWARD WARNER, SC. D.
ORVILLE WRIGHT, SC. D.
THEODORE P. WRIGHT, SC. D.

LANGLEY MEMORIAL AERONAUTICAL LABORATORY
LANGLEY FIELD, HAMPTON, VA.

AMES AERONAUTICAL LABORATORY
MOFFETT FIELD, CALIF.

AIRCRAFT ENGINE RESEARCH LABORATORY
CLEVELAND AIRPORT, CLEVELAND, OHIO

NATIONAL ADVISORY COMMITTEE
FOR AERONAUTICS

1500 NEW HAMPSHIRE AVE., DUPONT CIRCLE
WASHINGTON 25, D. C.

TELEPHONES: EXECUTIVE 3515, 3516, 3517

October 17, 1944.

Dr. Hsue-Shen Tsien,
California Institute of Technology,
Pasadena, California.

Dear Doctor Tsien:

With reference to my letter of September 23, there is forwarded herewith, by registered mail, for final check before release, a proof copy of the restricted Technical Note No. 961 entitled "The 'Limiting Line' in Mixed Subsonic and Supersonic Flow of Compressible Fluids," of which you are author.

It is requested that any necessary changes or corrections be indicated in red and that desirable changes or corrections be indicated in black in the enclosed copy of the Technical Note.

As it is desired to release this Technical Note in the very near future, I would appreciate your returning the corrected copy, together with the enclosed original material, as promptly as possible.

Very truly yours,

NATIONAL ADVISORY COMMITTEE
FOR AERONAUTICS

G. W. Lewis

G. W. Lewis,
Director of
Aeronautical Research.

Enc. Proof copy and orig. text and figs.

RESTRICTED

RESTRICTED

October 26, 1944.

Dr. G. W. Lewis
Director of Aeronautical Research
National Advisory Committee for Aeronautics
Washington, D. C.

Dear Dr. Lewis:

Following your request in the letter of October 17, 1944, I am returning the corrected proof copy and original manuscript of my paper entitled "The 'Limiting Line' in Mixed Subsonic and Supersonic Flow of Compressible Fluids". There are only two short corrections necessary and they appear on page 6 and page 24 of the proof copy.

It is indeed an honor to publish my paper as a technical note of the Committee and I deeply appreciate it.

Very truly yours,

Hsue-shen Tsien.

HST/pl
Encl.